AUTOMOBILE ENGINES FOR AIRPLANES

AUTOMOBILE ENGINES FOR AIRPLANES

JOE CHRISTY

TAB BOOKS Inc.
Blue Ridge Summit, PA 17214

FIRST EDITION
FIRST PRINTING

Library of Congress Cataloging in Publication Data

Christy, Joe.
Automobile engines for airplanes.

Includes index.
1. Airplanes—Motors. 2. Automobiles—Motors.
I. Title.
TL701.1.C49 1986 629.134'352 86-6013
ISBN 0-8306-2447-3 (pbk.)

Contents

Introduction

Amateur airplane builders have been fitting automobile engines to their creations for more than 50 years. Perhaps the best known among the early efforts were the Model A Ford engines which provided power for many pasture pilots in the Pietenpol Air Camper and Corben Super Ace, as well as many one-of-a-kind machines.

Several government-certificated airplanes were produced with car engines during the 1930s. The Funk cabin monoplane originally appeared with the Ford Model B engine, an improved version of the four-cylinder Model A. The Model B was produced concurrent with the new V-8 1932-1934. The Wiley Post, a biplane trainer built in Oklahoma City, was powered with a modified Model A engine. The low-wing Arrow Sport was equipped with a Ford V-8 in the mid-'30s, and the Mooney M-18 Mite was certificated with a 25-hp Crosley engine in 1947.

As good, used certificated aircraft engines in the 50 to 75-horsepower range grew scarce, homebuilt airplane constructors in the U.S. turned to Volkswagen conversions. Some experimenting with VWs for aircraft use in America began quite early. I saw one running on a test stand at a Tulsa air show in 1962. In any case, it was inevitable that the VWs would attract the homebuilt airplane people because these little engines are not only plentiful and relatively inexpensive, but are configured as lightplane engines—that is, four-stroke, four-cylinder, opposed-type, and air-cooled. True, they achieve their rated power at well over 4,000 rpm, which is too fast for a propeller of proper size, but that is true of all automobile engines, and one may either derate (turn it slower and give up some power), incorporate a propeller reduction drive (which adds weight, cost, and something else to break), or cut both rpm and prop diameter for a middle ground compromise at something over 3,000 rpm.

There are four principal VW systems. John Monnett, on Wittman Airfield in Oshkosh, Wisconsin, offers conversion parts for the simplest and lowest-cost airborne VW. Joe Horvath's Revmaster company in Hesperia, California, has led the way in upgrading the VW's muscle with larger cylinders, supercharging, and variable-pitch propellers. Rex Taylor's HAPI Engines company is located on the municipal airport in Eloy, Arizona, and Rex has concentrated on achieving maximum dependability with his VWs by mating the best VW components from several models and sources with such improvements as hydraulic valve lifters. And then there are the Limbachs, which are built in Germany and are "based on the VW." The Limbachs are certified by the West German equivalent of the FAA and power most European motorgliders. The Limbachs, in several versions, are available in the U.S. only as complete, ready-to-fly units from U.S. distributor J.C. Treager, who has an office in the Tulsa (OK) Municipal Airport building.

The Monnett conversion requires that one provide his/her own engine and convert with Monnett parts and instructions. One may do the same with HAPI or Revmaster or purchase a complete engine from either.

There is, of course, a great need for larger engines. Although commercially-produced airplane engines are available in the 100 to 200-horsepower range, there is great interest in converted automobile engines of this size because of the very high cost of certificated airplane powerplants. The 108/115-hp Lycoming 0-235 has a list price of $9,050 at this writing, while the 180-hp 0-360 is $11,295. This seems especially onerous when one considers that mass-produced automobile engines may be of superior quality.

During the fall of 1985, I spent some time in Wichita with Dave Blanton, whose Javelin Aircraft has for many years been a respected general aviation engineering firm, and Dave was flying some Ford automobile engines in two-place and four-place Cessnas that will get your attention. These are an in-line four cylinder (Escort and Lynx), and the 232-cubic inch V-6, respectively. We will investigate these engines in the following pages.

I also hied myself to Eloy, Arizona, and Rex Taylor's HAPI

Engines facility, where I saw the HAPIs carefully put together and absorbed many words of engine wisdom from Rex and son Pat. We will thoroughly examine the HAPIs momentarily.

My next stop was Duncan Aviation in Duncan, Oklahoma, where personable Dan Duncan and his crew were building a Bede BD-5 pusher to be fitted with one of Duncan's Wankel (Mazda) rotary engines. The Bede was on hold, however, while Duncan's crew reworked a Piper Tri-Pacer into Pacer configuration (tailwheel) and extended the nose to accept a rotary engine. These engines would seem to possess a lot of promise for aircraft use, but their gestation period has been a long one.

Meanwhile, I learned that Porsche had developed a lightplane engine and were flight-testing it in a Mooney 231 and a couple of European airplanes. However, I was told that, if produced, it would be priced *above* Lycomings and Continentals of similar power. That being the case, I fail to see why Porsche has bothered with such a project. The Continentals and Lycomings are outrageously priced—which is a principal reason that a book such as this will find so many readers.

Chapter 1

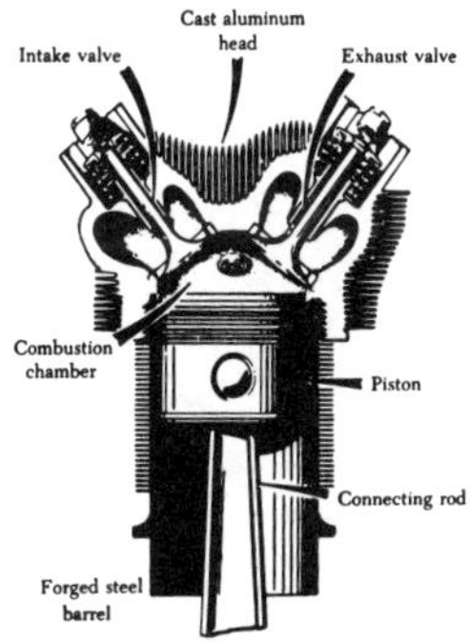

Where We've Been

Amateur airplane builders—and a few commercial builders—have been putting automobile engines in airplanes since the 1920s. To call any such installation "successful" depends upon one's definition of that term. At first glance, the Model A Ford engine seems to have served well, not only in a host of homebuilts, but in the ATCed Wiley Post. However, the success of the Model A conversion was owed to the times—this automobile was produced during the Great Depression—and to the fact that we didn't expect too much then. Also, the Pietenpols and other Model A-powered craft were most numerous in the Midwest and over the Plains, where there were plenty of suitable pastures to serve as emergency landing fields.

Ford Model A Conversions

The Model A Ford was produced 1928-1931 inclusive (slightly modified as the B and C models, fitted mostly to trucks, this engine was produced into 1934). Meanwhile, the first flathead Ford V-8 was introduced late in 1931 as the 1932 model. We'll come back to it momentarily.

Although the Model A Ford engine powered an uncounted number of one-of-a-kind homebuilts, it was primarily identified with the Pietenpol Air Camper. The first Air Camper took to the air in 1929, the designer, Bernard Pietenpol, told me. That was about the time that a new magazine, *Modern Mechanics & Inventions*, began publishing plans for homebuilt airplanes such as the Heath Parasol, Gere Sport Biplane, and others. Pietenpol, who had built several planes prior to the Air Camper design, told the magazine about his airplane and that led to the appearance of Air Camper plans in 1930 issues of that publication.

The Pietenpol was an immediate success. Despite its six-pounds-per-horsepower engine installation, it was hard to beat for simplicity or economy. Many were built from the magazine plans. Others were constructed from kits marketed by Pietenpol which sold for $63, with the builder furnishing his own landing gear, struts, dope-fabric, engine, and propeller. A complete airplane, built in the Pietenpol shop at Cherry Grove, Minnesota, sold for $750 fly-away-factory. The airframe was wood and fabric, with a special wing airfoil designed by Pietenpol. This airfoil was similar to the Eiffel-36 curve used in the Curtiss Jenny wing, and appears to have been one of the secrets of the Pietenpol's success. The Camper weighed 625 pounds empty and landed at 35 to 40 mph; rate of climb was around 200 fpm, cruising speed was 65 to 75 mph (depending upon the builder), and range was 200 miles with 10 gallons of "regular" automobile gasoline.

Pietenpol's conversion of the Model A engine involved a minimum of changes. It was turned around and the propeller attached to the flywheel flange by means of two hub plates, one of which had a pair of holes bored into it to fit the flange dowels. The valve cover casting was replaced by a lighter one made of sheet metal, as was the timing gear cover. The single magneto was mounted on a shelf at the timing gear end of the engine, and was driven by a coupling which screwed into a hole in the crankshaft originally meant for the hand-crank ratchet. The oil system was modified, and the oil pan redone in aluminum. Total weight, including cooling system, was close to 250 pounds (Fig. 1-1).

Pietenpol recommended that the Model A conversion be cruised at 1,600 rpm, which produced about 40 hp. The Model A torque curve peaked at 1,000 rpm. Pietenpol's propeller was WWI-surplus, 60 inches in diameter, with a 46-inch pitch. The Model A had a bore of 3 7/8 inches, a stroke of 4 1/8 inches, and a displacement of 200.5 cu/in. Compression ratio was 4.5:1. As late as 1966 one could buy a Model A rebuild for $160.

Other Ford A conversions followed the Pietenpol, some so extensively modified they were hardly recognizable, including the air-cooled Douglas of 1935. The Straughan/ Wiley Post biplane trainer received an Approved Type Certificate early in 1935 from the Bureau of Air Commerce (forerunner of the CAA/FAA). Priced at $990, this 600-pound (empty) open-cockpit two-placer (side-by-side) was first built in Wichita as the Straughan Model A; it became the Wiley Post when production was moved to Oklahoma City. At least two dozen were built (Fig. 1-2). More than 100 Ford-powered Akron-Funks were built just prior to WWII fitted with an extensively-modified Model B engine. Mounted inverted, this powerplant was said to develop 63 hp at 2,150 rpm. The Funk resembled a fat Piper Cub and sold for $1,950. See Figs. 1-3 through 1-5 for other interesting auto/aero powerplants.

Ford V-8 Conversion

The Arrow Sport F, manufactured in Lincoln, Nebraska, 1935-1938 inclusive, was powered with the flathead Ford V-8 rated at 85 hp at 3,400 rpm. A propeller reduction unit with a ratio of 1.95:1 kept propeller speeds at 1,700 rpm or less. This machine cruised at 90 mph, and seated two in the open cockpit side-by-side. Slightly over 100 were built and sold for $1,500. I took photos of a restored Arrow Sport at an antique airplane convention in Iowa some years ago, but blew the opportunity to get details of this engine conversion (Fig. 1-6).

Chevrolet Corvair

Bernard Pietenpol was still flying one of his famous Parasol monoplanes in the mid-1960s when I talked with him. This airplane was fitted with a converted Chevrolet Corvair engine that weighed 260 pounds ready to fly, including battery. Mr. Pietenpol seemed quite pleased with this installation, saying that he was particularly impressed by its lack of vibration. He went on to explain that, as in the case of the much earlier Ford conversion, he sought to adapt a car engine by making a minimum of changes. The vacuum control on the ignition system was removed in the interest of improved idling. No change was made in the distributor or its governor. The tachometer was driven from the bottom of the oil pump. An intake manifold was fabricated with a jacket around it to warm the intakes and help cool the oil. The cooling system was used just about as it was in the automobile; power taken by the fan was so small that it wasn't noticeable on the tach, so the fan was kept. He said that no heating problems had been encountered. The propeller was secured by threading the crankshaft on the inside for a one-inch shaft to take the prop load. Mr. Pietenpol told me that the Corvair was

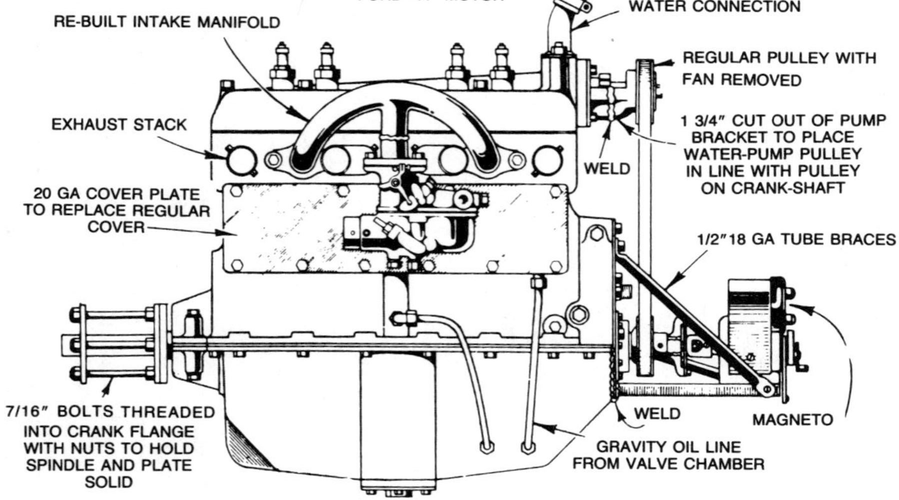

SIDE VIEW OF THE COMPLETED CONVERSION OF THE FORD "A" MOTOR

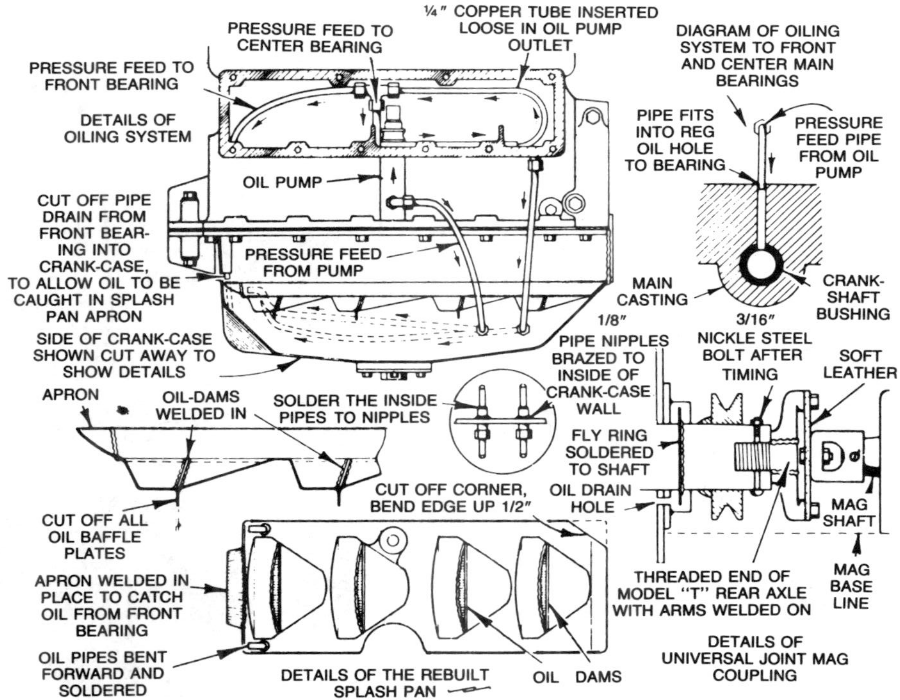

Fig. 1-1. The Model A Ford engine converted to airplane use produced 40 hp and weighed up to 240 pounds. Magneto coupling was "soft leather." Normal cruise at 1,600 rpm resulted in 35 hp. (courtesy 1932 *Flying & Gliding Manual*; Fawcett Publications, via EAA)

Fig. 1-2. The Wiley Post trainer (originally, Straughan) was side-by-side two-placer ATCed with a Converted Model A Ford engine. Built in Oklahoma City under the direction of pioneer flier Burrell Tibbs in the mid-'30s, the last of about 30 built were priced at $1,692. (courtesy Joseph P. Juptner)

rated at 85 hp, but he estimated that he was getting about 60 hp at 2,600 rpm, direct drive (Fig. 1-7).

Early Volkswagen Conversions

The 1200cc Volkswagen engine, which powered the original

Fig 1-3. Even the lowly Ford Model T engine was converted for flight by Bernard H. Pietenpol for his single-place Sky Scout in 1930. (courtesy B.H. Pietenpol)

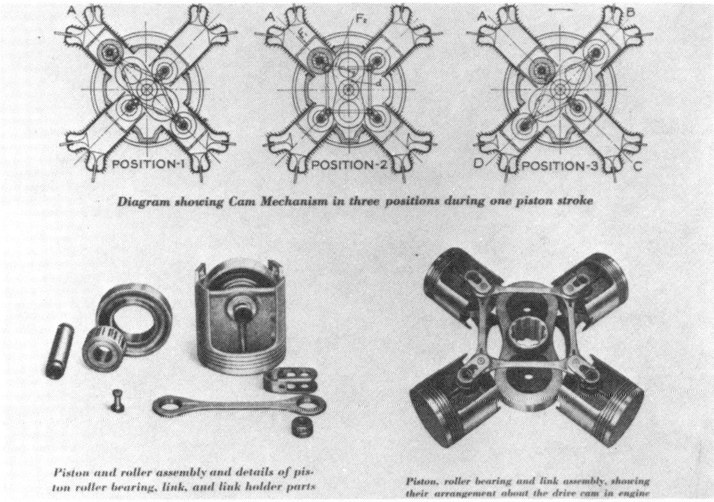

Fig. 1-4. The Duncan Wankel (Mazda) discussed in Chapter 4 draws attention to another aircraft engine of a unique design, the Fairchild-Caminez, which was ATCed in the late '20s. A roller bearing inside each piston rode directly on the large driveshaft cam. (courtesy *Jane's All the World's Aircraft*, 1928)

German "Bug," was fitted to European homebuilts such as the Jodel and Turbulent during the '50s—before the little automobile had begun to show up on U.S. highways (the VW "Peoples' Car" was actually designed before WWII). However, homebuilt airplane peo-

Fig. 1-5. The four-cylinder Caminez of 135 hp as installed in the Waco 10. The slow-turning Caminez allowed large diameter propeller. The engine was not developed after Caminez' death in a plane crash. (courtesy Charles W. Collection)

Fig. 1-6. The pre-WWII flathead Ford V-8 engine of 85 hp was ATCed as power for the two-place (side-by-side) Arrow Sport F built in Lincoln, Nebraska during the middle '30s. Priced at $1,500, more than 100 were built. Propeller reduction ratio was 2:1.

Fig. 1-7. Several homebuilt airplane constructors turned to the Chevrolet Corvair air-cooled engine during the '60s, most flying it in almost stock configuration. Piper experimented with it but abandoned the project due to the unavailability of parts. This one is mounted in a Pietenpol Air Camper. (courtesy B.H. Pietenpol)

ple in the U.S. and Canada were aware of this engine's potential from the outset. Bob Huggins, of Tulsa, OK, had VW conversions running as early as 1962. By 1967, the 1600cc VW was producing 50 hp at 3,200 rpm, ten million VWs had been built, and amateur plane builders in North America were taking serious notice of this little engine. Hugh Beckham conducted a popular VW forum at the EAA convention in Rockford, Illinois, that year, and Beckham, Huggins, and other major experimenters were putting considerable effort into the development of propeller reduction drives.

The three most obvious types of propeller reduction drives were designed for the VW: belts, spur gears, and planetary gears. The V-belt drives were the most popular initially, but torsional vibration—little understood by most homebuilders—took its toll, manifesting itself in stretched, broken, and slipping V-belts, along with chewed-up gear teeth in the other types. It seemed clear that direct drive was the more practical way to go with the VW conversion. The VWs turn well over 4,000 rpm to produce rated power, so the reduction drives appeared necessary to keep prop speeds reasonable and efficient. However, as the VWs increased in displacement, it became more feasible to derate (limit engine rpm and give up some horsepower) and go to direct drive.

Monnett VW Conversions

In the '70s, John Monnett and Greg Erikson worked together to develop a simple and inexpensive VW conversion. The Monnett conversions were based on stock (new) 1600cc and 1700cc engines, which have 88mm cylinders that bolt on the Type III cases with no machining required (Fig. 1-8). John developed a high-lift camshaft, new intake system, ignition, and propeller hub. Internally, the Monnett approach leaves the engine almost pure Volkswagen. "None of our engines are babied; we fly them pretty hard," Greg Erikson told me. "Our normal cross-country cruise is 3,200 to 3,300 rpm, although we often pick it up to 3,500 or 3,600 rpm. While the engines will withstand continuous operation at 3,700 or 3,800 rpm, we have set 3,600 rpm as our maximum continuous because of fuel flow . . .

"Some people worry that our VWs turn too high for propeller efficiency, and therefore the VW should be geared down for aircraft use. Well, as long as tip speed is kept below 950 ft/sec, there's no problem. Assuming a continuous operation at 3,900 rpm, our prop could be 54 inches in diameter before the tips start to go su-

Fig. 1-8. A VW converted to aircraft use with Posa carburetor and Monnett Y intakes, along with Monnett propeller hub and X engine mount and magneto. (courtesy Monnett Experimental Aircraft)

personic. But we use only 50 and 52-inch props. Remember, the power output is directly related to rpm. That is why we use the little propellers and allow the engines to turn up. These engines don't produce their rated power until they turn 4,000 or 4,100 rpm. So when we cruise them at 3,300 rpm we're pulling less than 50 hp. At normal cruise we burn four gallons per hour.

"Operating temperature and pressures are similar to those in aircraft engines, except for oil. Oil temperatures of 220 and 230 degrees F are not uncommon; if that seems high, remember that these engines commonly run around 275 degrees F in the VW busses and squarebacks at 60 mph on hot days. The use of an oil cooler will easily cut this down to 190 degrees F, but do not run the oil cooler than 170 degrees F.

"Oil pressure may also be something other than what you are accustomed to in regular aircraft engines. It must be remembered that the Volkswagen is a low-pressure system, and the standard oil pump supplies only 15 to 20 psi. With the high-volume pumps used on most aircraft conversions, it's not uncommon to see oil pressures of 60 psi. At idle, don't worry if this system goes below 10 psi. But use only aircraft oil—40 weight in winter, 50 weight in summer. Do not use automotive oil, especially the multi-viscosity types, in summer.

"We baffle cylinder head temperatures to run like normal aircraft engines. In the summer, it's not unusual to see 425 degrees F on climbout. Once leveled off in cruise, it should come down to 350 to 375 degrees F. The redline is 450 degrees F.

"We have exhaust gas temperature gauges (EGT) on all our planes. Mainly, we refer to these to set up the mixture adjustments on the Posa carburetors on installation, and then to monitor them for signs of change, since we do not have adjustable mixture controls (later versions of the Posa are available with mixture controls). At normal cruise, the EGT reads in the vicinity of 1,100 degrees F. Consistent readings near 1,500 degrees F indicate the need for a richer mixture. The denser air of winter in the northern U.S. usually calls for a richer mixture—say, one or two turns of the needle."

Monnett Propeller Hub

When Monnett set out to achieve a practical VW conversion at low cost back in the mid-'70s, John established the following guidelines:

- ☐ The engine must remain stock, and should have no modifications to the basic parts.
- ☐ A bolt-on propeller hub extension must be added.
- ☐ The engine should have a vibration mount similar to that on standard airplanes.
- ☐ The magneto should be easily installed, and positioned so that a streamlined cowl may be fashioned.
- ☐ The prices of the units should be within reason to provide an inexpensive conversion of the VW.

Monnett concentrated on the second goal above. A key part of the process of converting the VW engine to aircraft use is to devise a way to attach a propeller. The early conversions required that the engines be disassembled and a three-degree taper machined to the pulley end of the crankshaft. Then, a steel hub was machined to be pressed onto the crankshaft. An alternate method was to merely shrink-fit (by heating) a hub onto the pulley end of the crankshaft. In both cases the standard VW keyway was used. Neither of these methods was strong enough to take high G-loads when spool-type extensions were employed, but these methods were acceptable at the time for the kind of aircraft for which they were used. The prop hubs required wooden spacers to enable the propellers to clear the left-front exhaust stack. Those builders who wanted a little more clearance for better propeller efficiency or improved streamlining would machine a three-inch or so prop-hub extension spool.

You have probably anticipated the problem that these methods incurred. With the extra extension moving the propeller outward, it was, in effect, making the lever arm longer, and transferring all the gyroscopic prop loads onto the crankshaft—and in magnified amounts. The spinning propeller is a big gyroscope; it wants to remain in one plane. Any turns, climbs, or descents result in increased loads on the crankshaft via the prop hub as the propeller (read: "gyroscope") tries to resist the change in plane. The more sudden the change or movement, the harder the propeller resists, and the more Gs placed on the crankshaft. The farther out the propeller is moved, the more the loads are magnified.

Monnett designed two such units, their "Super V" propeller extension unit which incorporated a four-inch housing and shaft plus a sealed, heavy-duty bearing; and a simpler "E-Cono-Vee" short hub prop mount. This latter component is a shrink-fit 4140 steel hub that is long enough so as not to require any extra block or extension to allow the prop to clear the exhaust stack, yet not so long as to significantly magnify the gyroscopic loads on the crankshaft. This unit is available from Monnett, along with the Monnett intake system and other conversion parts.

The Super V unit drawings are available.

Today, Monnett also continues to stock the oil bypass block-off casting and the fuel pump block-off casting. The VW engine, as used in the car, has an oil cooler mounted over the number three cylinder. This has been a cause of burned exhaust valves due to the restricted airflow over that cylinder. The oil bypass casting allows the oil cooler to be mounted elsewhere in the engine compartment. A casting is also available which allows the oil cooler to be mounted directly on top of the engine.

Magneto Installation

The early VW conversions used either some form of distributor system for ignition, or a large magneto that was designed to operate from the original distributor hole. With such installations any kind of a cowling was nearly impossible to design, because the magneto protruded above the engine three or four inches. Another early practice was the mounting of the VW engine directly to the airframe, with no attempt to reduce vibration. Obviously, that had to be changed.

Actually, the problem of mounting the magneto was largely solved before Monnett began its development program, but fabrication of a reliable magneto coupling was something else. Streamlining the magneto mount was accomplished by switching to an aircraft-type magneto and mounting it at the rear of the engine. That meant that a mag drive had to be designed that attached to the end of the crankshaft, along with a mag coupling. The coupling was intended to take up any small misalignment between magneto and crankshaft, and absorb the starting impulse of the mag. This was a troublesome component for a time.

Monnett made mag couplings from various kinds of steel, rubber, nylon, phenolic, you name it—all without much luck. The two main problems were the tremendous beating the coupling had to take from the impulse when starting, and the heat buildup and radial loads during operation. The steel drives were too strong, causing the dogs on the mag or the drive to fail. The rubber couplings distorted and cracked from the impulse, and became oval-shaped from the radial loads. The phenolic ones cracked into many pieces, the

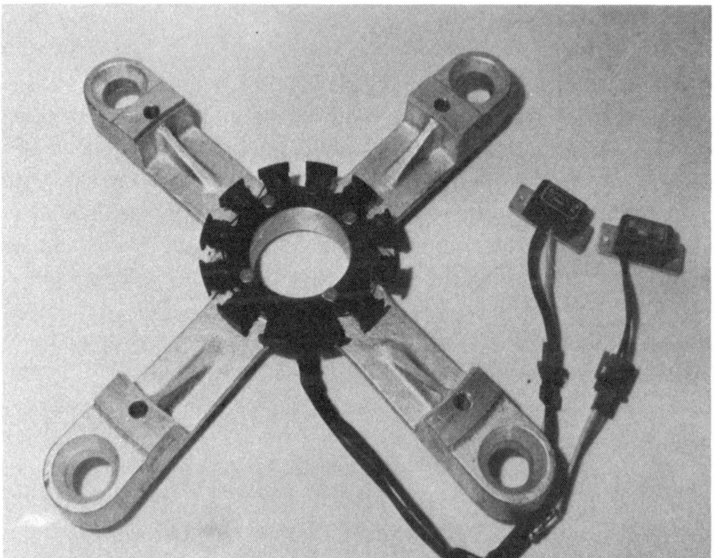

Fig. 1-9. The Monnett engine mount accommodated alternator and magneto. (courtesy Monnett Experimental Aircraft)

early aluminum ones were easily distorted by the impulse, and the nylon ones melted into bubblegum consistency.

Monnett at last settled upon 2024 T3 aluminum, and along the way the drive piece was redesigned to obtain proper lubrication and ease of replacement. This magneto drive was a machined piece that duplicated the center of the flywheel and bolted onto the crankshaft with the standard VW gland nut. The inside thrust surface was duplicated, and all thrust loads taken as in the car. To solve an oil leak problem, the drive was machined to accept the standard VW O-ring crankshaft seal.

While designing a better way to hold the magneto in place, Monnett decided to incorporate it into an engine mounting that contained vibration dampeners. That resulted in an almag casting in the form of an X, with webs for strength. This development was later adopted in principle by the other VW engine converters.

Monnett uses the Slick 4016 magneto. It is made for the Volkswagen, weighs but three pounds, and has proven very reliable. This system uses but one magneto. The extra mag, with its necessary wires and plugs, would add 10 pounds and Monnett feels that an additional spark plug hole weakens the cylinder heads and promotes cracking. HAPI and Revmaster, of course, disagree.

Carburetion

The original Solex and Bendix-Zenith carburetors were discarded by Monnett because of their complexity. There were high-speed and idle jets to adjust, complicated linkages to set up and align with the throttle control, and there was the fact that, with a float set-up, the engine would bog and die at high G-loadings. Also, there was the problem of maintaining a proper and constant fuel pressure without using the engine-drive fuel pump. The fuel pump was one of the original VW accessories discarded by Monnett at the beginning of the conversion process. They never considered an electronic fuel injection system because of the cost and complexity. They insisted on keeping things as simple and foolproof as possible.

The answer to Monnett's carburetion problem was the simple, lightweight, and inexpensive Posa Injector Carburetor. Actually, the Posa is neither a carburetor nor an injector; it is simply a device that measures and mixes fuel and air. The principle isn't new, as the motorcycle and radio-controlled model airplane people have used similar devices for years.

The operation of the Posa is so simple it leaves many people skeptical. It has but one moving part (in its simplest model), the slide. Screwed into the slide is a long needle with a ramp machined on one side. The slide has a hole in one end. As the throttle is opened, the slide and needed are moved. Since the needle has the ramp on it, as it slides out of the fuel inlet (needle orifice in Fig. 3-10), it allows more fuel to flow into the throat. Also, as the slide moves, the circle opening becomes greater (as looking directly into the air inlet), allowing more air to enter. It is here that the fuel and air are mixed. Basically, the system works on demand; i.e., the more the slide is opened, the more air is allowed to flow through, and the more fuel is allowed to enter. The airflow actually draws fuel from the line. Interestingly, Monnett advises against ram air to a Posa, while HAPI recommends it.

Intake Manifold

The Monnett conversion system provides a Y casting that mounts to the lower rear of the engine case to mount the carburetor. Metal tubing, with 90-degree elbows, connect with 1 1/2 inch ID rubber hose and lead to the intake head castings on each side.

Monnett's address is:

Fig. 1-10. The 1947 Mooney M-18 Mite was a single-placer, fitted with the 26-hp Crosley automobile engine. Originally produced in Wichita, Mooney built the last Mites in Kerrville, Texas, in 1953. The price was $1,995 with the Crosley engine, but most Mites were later powered by 65-hp Lycomings and Continentals. (courtesy Joseph P. Juptner)

Monnett Experimental Aircraft
P.O. Box 2984,
Oshkosh, WI 54901
Phone: (414) 426-1212

Crosley-Powered Mooney Mite

We should note the Crosley engine conversion that was fitted to the original Mooney M-18 Mite (Fig. 1-10). Although Al Mooney had designed a lot of famous airplanes (beginning with the Alexander Eaglerock in the mid-'20s), the Mite was the first to carry his name. It appeared in 1947 and was granted an Approved Type Certificate powered with the 26-hp Crosley "Cobra." Priced at $1,995, the M-18 was a single-place low-winger, largely covered with plywood, and had a range of 400 miles with its eight-gallon fuel supply. The Mite cruised at 85 mph, had an initial climb rate of 450 fpm, a 12,000-foot ceiling, and a power-off, flaps-down stall of 40 mph. Its propeller reduction unit was turned by multiple V-belts. Interestingly, Dave Blanton of Javelin Aircraft worked on the Crosley engine conversion. Dave was with Boeing-Wichita then. I was unable to learn just how many Crosley-powered Mites were produced, but 26 hp clearly wasn't enough and the Mite soon went to the 0-145 Lycoming, and ended up after five years in production as the M-18C, fitted with the 65-hp Continental.

Chapter 2

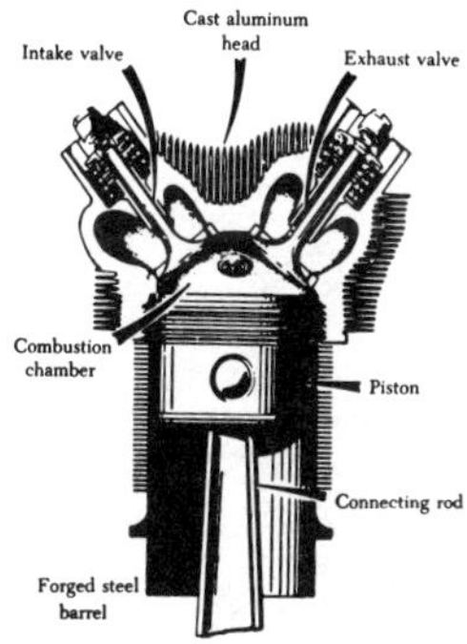

The Javelin Ford Engines

Dave Blanton is a former Boeing engineer who founded his own company, Javelin Aircraft, in 1953. He has contracted a number of special projects for the Wichita aircraft manufacturers, and for Aero Commander down in Betheny, Oklahoma—clearly, a somewhat more profitable activity than serving as a full-time employee of one of the commercial airframe builders (Fig. 2-1).

Dave had recently moved from Wichita into a brand new facility on the Augusta Airport (about eight miles east of the Beechcraft plant on Kellog Avenue) when I visited him just before Christmas, 1985.

The weather was cold with fog and intermittent near-freezing rain, below minimums for hundreds of miles in all directions. Obviously, we weren't going to fly the Cessna 175 Skylark Dave has used as a test bed for the Ford V-6 engine, but Dave started it, just as you would an automobile, and ran it for about 10 minutes while he explained the test gauges, especially the coolant pressure gauge and the fuel flow meter. It is a quiet engine, rated at 260 hp at 4,800 rpm (2,400 propeller rpm). It is cruised at 4,000 rpm (2,000 prop rpm), where it produces 132 horsepower burning 6.8 gallons per hour (GPH) of automobile unleaded supreme gasoline. That provides an indicated airspeed of 120 mph on the test airplane at 3,000 feet MSL and 60 degrees F. Manifold pressure is 19″ Hg at this setting. Blanton does not recommend turbocharging.

Fig. 2-1. Dave Blanton began his automobile engine conversion tests with the 140 cu/in (2.3 liter) Ford Mustang engine in 1976. Flown with a turbocharger, fuel consumption was excessive at 23 gph. Engine produced 230 hp at 5,400 rpm and 46 inches Hg. Reduction drive was 2:1. (courtesy Javelin Aircraft)

Development

The Javelin Ford engines are the result of more than five years of testing with the V-6, and with the several four-cylinder Escort/Lynx engines (Fig. 2-2)—plus some preliminary work on the Ford 2.3 turbo out of a 1979 Mustang Cobra. After the V-6 appeared, the Cobra engine was quickly abandoned because the 230

V-6 has 92 cubic inches more displacement while being 50 pounds lighter. The turbo burned 22 GPH and the V-6 burns 15 GPH, developing 231 and 260 hp respectively.

While the Javelin Ford 230 V-6 appears to be a viable alternative to the Lycoming 0-360 in homebuilt aircraft, the newer (deliveries began in the fall of 1984) Ford 1.9 four-cylinder engine should have the widest appeal because it is in the same power class as the Continental 0-200 and the Lycoming 0-235—which is to say, 100 to 120 hp. Javelin calls this one its Model 116 and, although it will produce more than 150 hp fitted with the Javelin intake manifold and roller camfollowers, Blanton prefers to turn it at lower rpm and describe it as 100 hp (Figs. 2-3 through 2-6).

The 116 weighs the same as an 0-200 at 198 pounds dry. The reduction drive adds 67 pounds for a total of 265 pounds, again, about the same as an 0-200 with all accessories, baffles, and exhaust.

Dave Blanton and his crew invested a lot of effort in the development of the earlier Escort/Lynx as an aircraft powerplant. This was the 1.6 liter engine, or 98 cubic inch, so it was the Javelin Ford Model 98 (and 98T as a turbo). "We flew the Model 98 and 98T

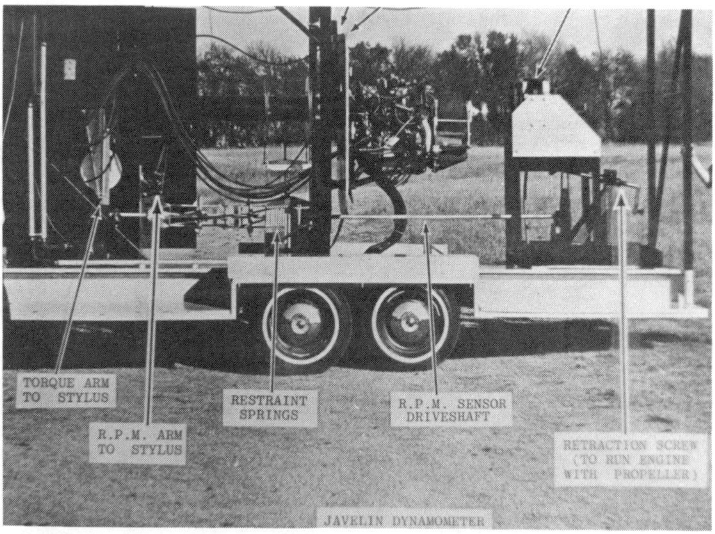

Fig. 2-2. Javelin Ford's dynamometer with Ford Escort/Lynx 98 cu/in engine in test. Later flown extensively in a Cessna 150, this engine conversion performed well, but proved heavier than the Continental 0-200 it replaced. The later 116 cu/in Escort is the current 0-200 "replacement." (courtesy Javelin Aircraft)

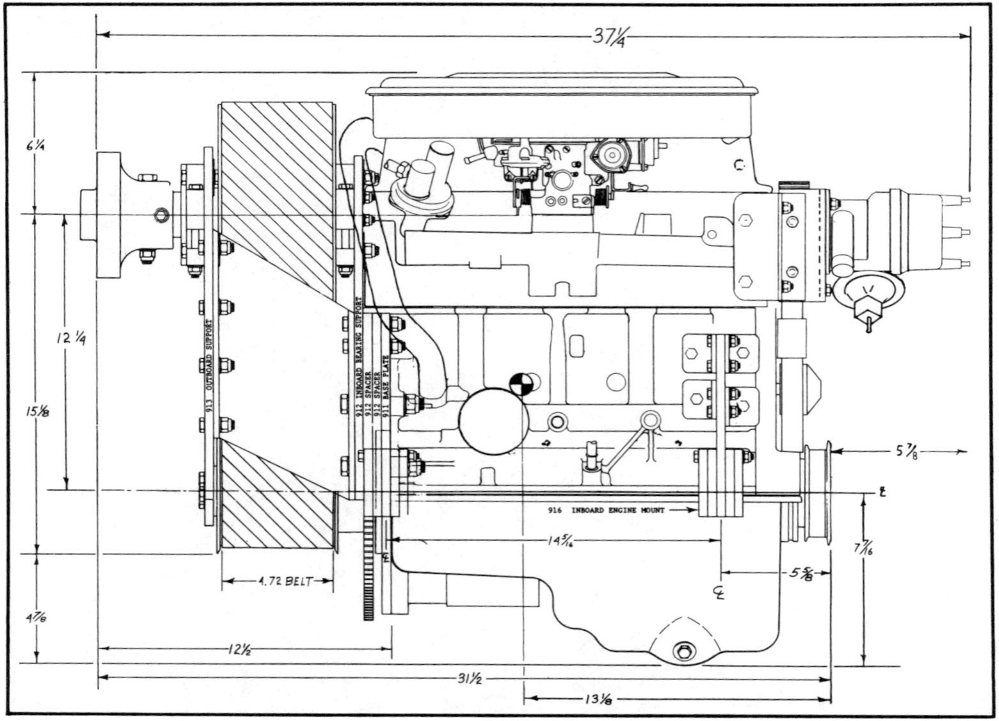

Fig. 2-3. Javelin Ford Model 116 dimensions.

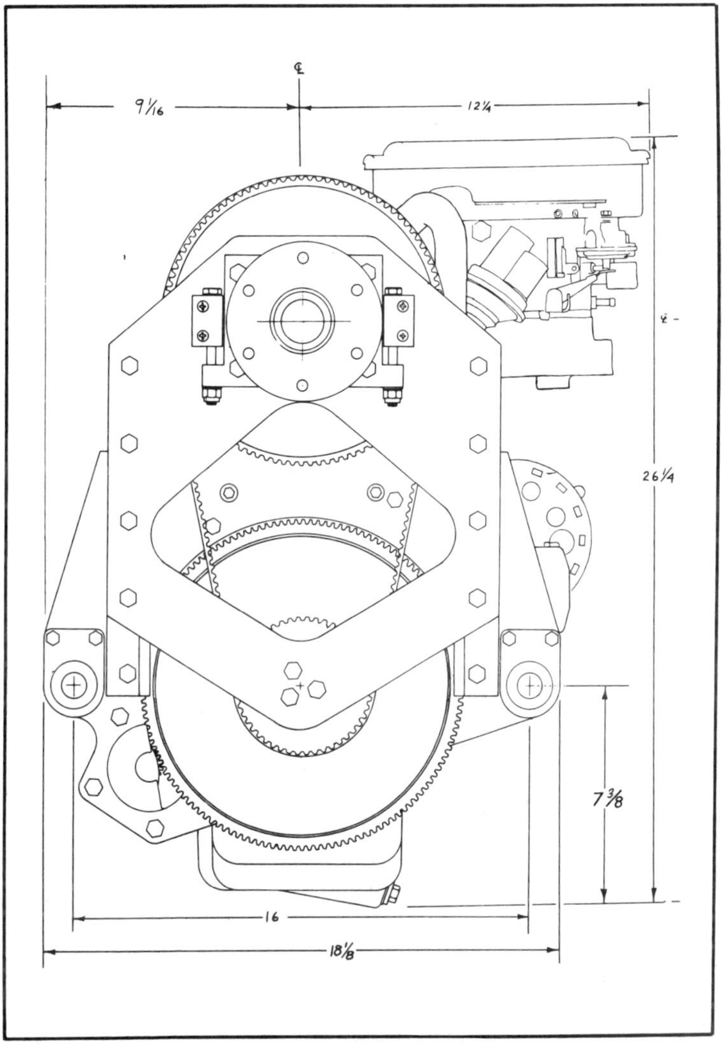

Fig. 2-4. The front support of the reduction drive is one of the secrets of the Javelin Ford 116's reliability.

for two years in our Cessna 150 test plane," Dave says, "and I can tell you now, the turbo is a mistake. The turbo version runs as smooth as a sewing machine. It was the smoothest and quietest engine anyone ever flew other than a turbine. But the reserve horsepower for climb wasn't there. Then Ford increased the size of the engine 20 percent and now we have that reserve. The turbo does not add enough power to make up for the weight increase. We are

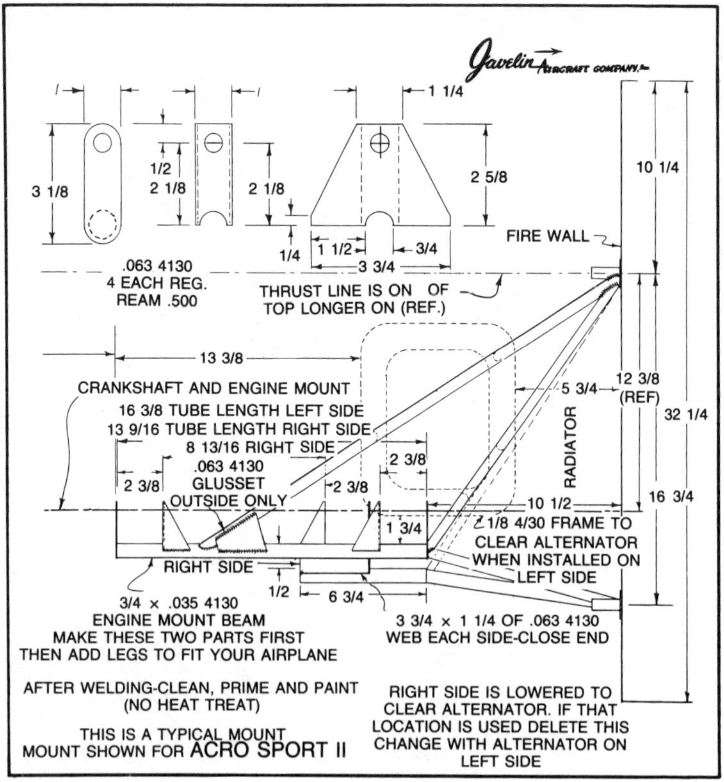

Fig. 2-5. The engine mount for the Javelin Ford Models 98 and 116 on the Acro Sport II homebuilt. Tubing is 3/4″ × .035 chrome-moly 4130.

leaving the model 98 or 1.6 liter data in the Javelin engine handbook to show that we have been there; we probably spent $75,000 on the Model 98 and 98T project, but now we tell our builders, 'DON'T do it.'

"We have run the Model 98 with the big intake manifold we developed, and the Holley 34-7448 carburetor, at 131 hp. The Model 116 (1.9 liter) with the same manifold, carburetor, and roller camfollowers will produce over 150 hp. We expect to develop an aircraft-type camshaft for the roller camfollowers, and we expect to install this engine in a Pietenpol Air Camper for test during the summer of 1986.

"The stock 1.6 and 1.9 engines come with 9.5:1 pistons (pistons with a 9.5:1 compression ratio). If you add the turbocharger, these pistons must be replaced with 8:1 pistons. At about 35 inches manifold pressure (Hg) with the turbo, the engines will develop as much power as with the higher compression 9.5:1 pistons. At 46 inches manifold pressure there is a 10 percent increase in horsepower, but the fuel flow must be increased considerably to keep the turbine inlet temperature down to 1600 degrees F. Meanwhile, the size of the radiator must be doubled with a turbocharger.

"The weight increase is not offset by the power increase. You will get more performance with 9.5:1 pistons and a tuned exhaust—but a muffler must be installed downstream. The turbocharger is a very effective muffler (and is lighter than a muffler); it makes the engine have a pleasant sound. But without the turbocharger, a muffler must be installed. The noise with a straight stack without a muffler will cause ear damage. I enjoy the sound of a nice-running engine, but this one without a muffler is not tolerable.

"These engines are rated by Ford at 6,300 rpm. In race cars they are turned up to 11,000 rpm. We cruise the engine at 5,200 rpm, which is 2,600 rpm on the propeller. In the automobiles, these engines demonstrate a normal time before major overhaul (TBO) of 3,500 hours. A TBO of 2,000 hours is reasonable."

These four-cylinder engines have overhead cams, aluminum cylinder heads, and cast iron blocks (Figs. 2-7 through 2-9). The cylinder head uses a compound valve hemispherical (CVH) design. The CVH describes the hemihead valve configuration where the

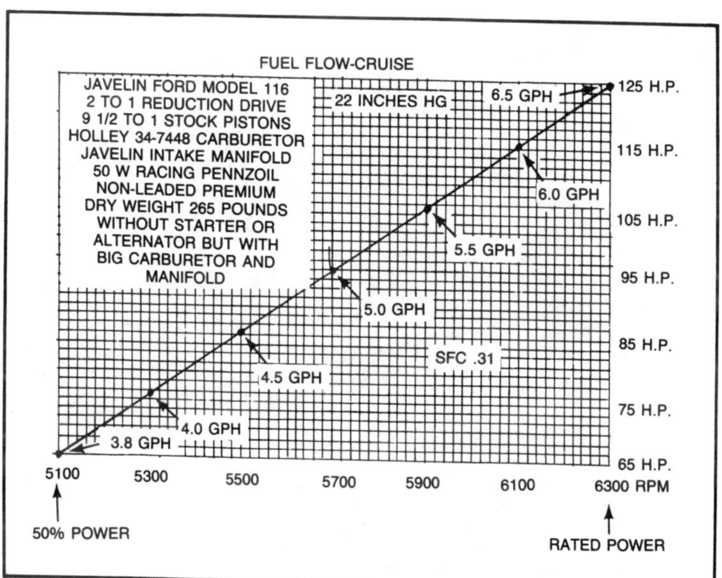

Fig. 2-6. Power, rpm, and fuel flow for the Model 116.

Fig. 2-7. Dave Blanton says that the girl model is to aid one in judging the relative sizes of the engine components on the Model 116 engine. Sure, Dave. (courtesy Javelin Aircraft)

valves are canted for improved breathing. A hemispherical combustion chamber allows larger valves to be used, and creates more horsepower with better fuel economy.

Turbocharging

One of the most valuable features of the Javelin engine manual is Blanton's willingness to detail development mistakes ("We're not salesmen, we're engineers," he says). Javelin's first automobile test engine was the 1979 model 140 cubic inch Cobra Turbo. "We stripped off the Ford automatic wastegate control," Dave says, "believing that we would control the manifold pressure with the throttle. We had the idea that we were going to operate at 48 inches Hg, and we flew that engine that way. Then we could not evaluate various camshafts. That's when we stopped the program and built our dynamometer. With the dynamometer we could quickly evaluate camshafts, but then we discovered that when we went from 42 inches manifold pressure to 48 inches, which we assumed would produce more power, all we were doing was making more heat. Then we put Ford's automatic wastegate back on to control the manifold pressure at 42 inches and got the throttle fully open; the power went from 180 to 231 hp. Don't make the same mistake. In a progressive carburetor you must get the secondary fully open to get maximum power.

"The hottest part in a four-cycle engine is the exhaust valves. The second hottest part is the exhaust valve guides, and the third hottest part is the dome of the piston. When you are going to operate an automobile engine at high power, the first thing to look into is the type piston in the engine. Any piston that has a slot for oil

Fig. 2-8. Javelin Ford Model 116; 295 pounds, 150 hp at 6,300 rpm. Cruise at 65 hp on 3.8 gph. With some two million of these engines on the nation's highways, good used ones are not hard to find. (courtesy Javelin Aircraft)

Fig. 2-9. Rear view of the Model 116 four-cylinder engine reveals accessory drives. (courtesy Javelin Aircraft)

return is going to suffer a melted dome with turbocharging. The early Escort piston of 8:1 compression had an oil return groove that cut the piston in two at least 40 percent.General Motors engines have what is called a 'T-slot' piston. But all aircraft pistons are solid trunk type to conduct the heat out of the dome into the skirt. The 9.5:1 pistons for the Escort are solid trunk, as are the new 8:1 pistons for the turbo version.

"Air-cooled engines are just barely able to cool at 1/2 hp per cubic inch; so what happens when you add a turbocharger? They may have the automatic wastegate set at 35 inches, or as low as 32 inches. The FAA certificated air-cooled engine cannot use the turbocharger to increase the horsepower at sea level. The turbocharger is used to hold rated power as the airplane climbs.

"On a liquid-cooled engine, the turbocharger can be used for a big increase in power, but you must cool the dome of the piston with fuel. The unsupercharged liquid-cooled engine can be operating at a specific fuel consumption (SFC) of .37, but you put on a turbo and you must put larger jets in the carburetor to hold down the turbine inlet temperature to a maximum of 1,650 degrees F. The resulting SFC will be .70; so when you double horsepower with a turbo, you also double fuel consumption at that power, which is four times what you had to start out with.

"Turbochargers are very popular on automobiles, but they are mostly just going along for the ride. The turbocharger is running at its normal rpm of 130,000 to 140,000, but the manifold pressure is being restricted to low pressure by the throttle valve. The average automobile will be cruising at 20 to 21 inches at 55 mph., and 65 mph at 22 inches. At full throttle for passing, 46 to 48 inches are available, which makes the car accelerate and go like hell.

"The manifold pressure available from the turbocharger is used for about five seconds at a time. Most of the new 1985 and 1986 automobiles have electronic fuel injection to meet the exhaust emission requirements throughout the power range. This system is made to order for a turbocharger since the engine can operate lean at low power and dump in excess fuel for a burst of power to cool the domes of the pistons.

"We have no need to control emissions throughout the power range on an aircraft engine, so we have no interest in the hazards of electronic fuel injection (if an injector nozzle gets plugged, the electronic fuel injection goes to automatic retard and you won't have enough power to remain airborne). But if we use a turbo for increased power we must jet the carburetor to keep the EGT down to an inlet temperature at the turbo to a maximum of 1,650 degrees F."

I asked Dave to explain the significance of specific fuel consumption.

"Well," he replied, "the lowest specific fuel consumption at which an engine will run is .31. That is .31 pounds of fuel per hour per horsepower." He paused to see if that was sinking in, then continued. "Any high-compression liquid-cooled engine will run at this

SFC. An air-cooled engine can be run at this SFC only at very low power. As power goes up, there is a tremendous increase in SFC with an air-cooled engine. They will go from .31 to .70 at rated power at only 1/2 hp per cubic inch, while the liquid-cooled engine can develop as much as 1.3 hp per cubic inch on an SFC of .37. We derated the V-6 engine from 5,400 normal rated rpm to 4,800 rpm for 260 hp, or 1.12 hp per cubic inch at an SFC of .35."

Reduction Drives

Blanton considers his Cogg belt reduction drive system to be totally reliable, so much so that he says he'd never consider any other drive system (Fig. 2-10). He estimates belt life at 2,000 hours.

"We have never replaced a belt," he says, "and we cannot see any sign of wear. There are a number of unsuccessful reduction drives that will destroy the belt in as little as 20 minutes. If a reduction drive with a Cogg belt does not have a front support, this means that it cannot be centered and the tension properly set. A one-fourth turn on our adjusting bolt will make the belt run full forward aft. A one-fourth turn is eleven thousandths. If there is no way to center the belt, and it is rubbing on a belt guide to keep it on the sprocket, belt life will be about 20 minutes. If you don't have the proper tension on the belt, you will feel a serious vibration that is going to fail something. The tension specified by the manufacturer is achieved by placing a steel bar against the belt in the center between the sprockets and applying a 50-pound load. The deflection should not be more than one-eighth of an inch. This is a tension load of 700 to 800 pounds. There is no way you can get proper belt tension without a front support.

"For the Model 98 engine reduction drive we very carefully designed it to 2.04:1 to break up any resonant frequencies that four pistons might find. This was a big mistake; the belt was changing one tooth with every rotation, and in each second it is searching out the entire spectrum. In flight, in the cockpit of the Cessna 150, the sound is yhing-yhing-yhing. We had 98 teeth on the top sprocket, and 48 on the bottom sprocket.

"We now know that any ratio near two-to-one, on either side, will cause a terrible searching sound as it goes through the entire spectrum every second. So, we consider any ratio from 1.8 to 2.2-to-1 as bad. We put 2.86:1 on an airboat and it is okay. We used a 14mm belt on the 140 cubic inch turbo Ford. It sounds like the supercharger whine on a Rolls-Royce Merlin.

Fig. 2-10. Front view of the Model 116 shows starter gears and the important reduction-drive adjustment. Blanton claims no belt failures during nine-year test programs. (courtesy Javelin Aircraft)

"The 8mm belt we used on the first 98T engine was run two years, with as many as 700 starts and runs, plus about 200 dynamometer runs where we threw on load to stall the engine. After all that abuse, there was no sign of wear on the belt. Meanwhile, with a 2:1 reduction drive the unit is perfectly quiet; doesn't make a sound. The same tooth comes down in the same groove every other rotation, and there is no searching.

"Of course, if there is a bad spot with the propeller you are using, you can re-index it. On most direct-drive engines the propeller is very carefully indexed to get the lowest vibration. This is so critical on some engines that they have a dowel pin to index the propeller. There are many metal propeller failures on direct-drive engines, and the situation is worse with geared reduction drives. These vibrations are gone with our belt drive."

Cooling

During the past six years Javelin Aircraft has flight-tested at least 20 different radiator configurations for the four-cylinder and V-6 Ford conversions. "We have flown over our airport a dozen times leaving a trail of steam," Blanton says ruefully. "In most cases we just did not have enough inches of tubes to cool at low pressure. . .

"We learned to watch the cooling system pressure gauge during takeoff and a circle of the field. A cooling system that is not going to cool will reach 16 pounds per square inch (psi) in three minutes. If it's going to cool, it will stop by nine psi; but if it goes past nine, we start getting into position to land. When the pressure gets to 16 psi, and we have a 16-pound pressure cap, we close the throttle and the flight is over.

"We are embarrassed to admit a false assumption we made concerning the installation of a cooling fan. We just knew that if everything else failed, that was a sure way to cool a liquid-cooled engine. We never flew the 140T engine with a cooling fan. We just assumed it would work, and we could not have believed anyone if they had told us it would not work. Every automobile on the road cools that way.

But when we put cooling fans on the V-6 they would not cool the engine. We tried pusher fans, puller fans, five-blade, seven-blade plastic, and the big five-blade steel fan used on Ford trucks; but on every flight test we would reach 16 psi in three minutes. We built beautiful heliarc-welded aluminum shrouds to make the fan work, but on every flight test the situation was unchanged. We tried reverse flow, pulling air in at the base of the windshield; it did not work. Finally, we got rid of the shroud and let the air blow through and out the bottom, and it cooled. The solution was a lot of inches of tubes in the radiator, and a good pressure differential. A factor in our need for a large cooling area is the heat exchanger we have in the oil system. It's all cooling, and it does not matter if it's oil or coolant.

"The radiator cores should always be in series, never in parallel. As the hot coolant moves forward it is progressively cooled until it leaves the front core at the lowest temperature. As many as four or five cores may be placed one behind the other to get the necessary number of tubes."

Cooling Drag

"Cooling drag is a very serious factor," Blanton continued. "The big square radiator we tried on the Cessna 175 was a complete failure, and the beautiful radiator on the belly of the Cessna with the Cessna 337 scoop just barely cooled the engine (Fig.2-11). The cooling drag with the radiator below the cabin cost an estimated 60 horsepower. Our last installation with the V-6 engine put the radiator behind the engine and ahead of the firewall. This has been effective. We need a lot of radiator on the 230 V-6 because we are running it as lean as possible, and because we are also cooling the oil. The Javelin manual details all of this and gives all of the numbers. It provides the core size and number of tubes needed, along with the proper (critical) height of the header tank.

"I think the oil cooling problems were solved for years to come by introduction of the Ford heat exchanger in May 1984. It not only

Fig. 2-11. Air Scoop on the belly of the V-6 powered Cessna 175 Skylark proved to be one of those good ideas that didn't work out. (courtesy Javelin Aircraft)

fits the four-cylinder engines discussed here, but the V-6 as well. We are proud that Javelin Aircraft pushed for this development for two years. Every time our poor little Cessna 150 sat in a hayfield with oil dripping off of it, we would scream so loud that everyone in Detroit could hear us. We think that our input was a contributing factor to Ford's development of this heat exchanger to replace the oil radiator.

"However, remember that this makes the cooling system more critical. You don't cool anything with a radiator hung outdoors because air will not go through it. Pressure differential is required to make air flow through the core. When we specify a given number of inches of tubes, that is what is needed with the heat exchanger. You should not fly these engines without the heat exchanger in the oil. It was good fortune that we put the radiator inside and got more cooling at the same time we put on the heat exchanger."

Roller Camfollowers

Dave Blanton is downright evangelical on the subject of roller camfollowers, and if his figures are correct, he has every right to be. The roller is built into the rocker arm on an overhead cam engine, and into the hydraulic lifter on a pushrod engine.

You cannot use a roller on a camshaft made for solid camfollowers. With rollers riding on the cam lobes, you need a camshaft made for that setup. Ford went to roller camfollowers on the 1985 1.9 liter engine (Javelin Ford Model 116), and Dave was happy about that; he says there is a gain of 15 hp with this improvement alone. At the time of my visit to Javelin Aircraft, Blanton did not have a V-6 with roller camfollowers. He expected the 1986 V-6 Fords to be so equipped.

Fuel and Oil

The Javelin Ford manual is very explicit about fuel and oil. Use only automobile non-leaded premium, 100 octane av-gas, or LL100 av-gas. These engines will be destroyed very quickly if operated on 80 octane av-gas or automotive gasohol containing methyl alcohol.

Use only 50W Racing Pennzoil without exception.

Weights

The GO-300 Continental of 175 hp removed from Javelin Aircraft's Cessna 175 weighed 391 pounds less propeller, spinner, and oil. The Javelin Ford 230 V-6, complete with all accessories, weighs 387 pounds. The Continental, while requiring 10 quarts of oil to the V-6's five quarts, ends up 24 pounds lighter after 37 pounds are added to the V-6 installation for radiator, coolant, and tail pipes.

The actual displacement of the Windsor V-6 (built at the Windsor, Ontario, Canada Ford plant) is 232 cubic inches. At a reduction drive ratio of 2:1, the swept volume per propeller rotation is 464 cubic inches, which compares with the 230-hp Continental O-470 in displacement and horsepower. The O-470, complete with all accessories, weighs 488 pounds dry. The V-6, at a total of 434 pounds, is 54 pounds lighter and, according to Blanton, a much higher quality engine (Figs. 2-12 through 2-19).

The Javelin Ford V-6, complete with cooling system, weighs 47 pounds more than the Lycoming O-360 of 180-200 hp. However, the O-360 will burn 20 pounds more fuel per hour (and the O-470 will require up to 36 more pounds of fuel per hour).

The Javelin Ford Model 116 four-cylinder engine (1.9 liters) weighs 265 pounds without starter and alternator; 295 pounds with all accessories. Fitted with the Javelin intake manifold and Holley carburetor, this engine will produce 95 hp at 5,700 rpm (2,850 propeller rpm). Turbocharging is not recommended, as previously discussed.

Fig. 2-12. The compact Javelin Ford 230 cu/in V-6 has been developed for aircraft use with propeller reduction ratios of 2.82:1, 2:1, and 1.61:1 as required by type of installation. The T-18, Glassair, and Midget Mustang use 1.16:1, while replica fighters with four-blade props use 2.82:1 ratio. (courtesy Javelin Aircraft)

Fig. 2-13. The quiet V-6 produces 260 hp at 4,500 rpm (2,400 propeller rpm). Specific fuel consumption is .31, while .41 is the best an air-cooled engine can do. After the above photo was made, the scoop was removed from the plane's belly and radiator mounted behind the engine inside cowling where cooling is most effective. (courtesy Javelin Aircraft)

To Oshkosh with the V-6

When I asked Dave about his trip to the EAA Fly-In at Oshkosh in 1985, he visibly brightened. "That was a blast," he said. "Four of us got into the Cessna 175, and we flew in formation with three other airplanes. With the MFC-8467 prop, we only needed to turn it at 1,900 rpm to stay with the group. That was 50 percent power, or 132 hp. At most of the stops we would still be over half full of fuel, and the other airplanes would always refuel, but we would just tell them that we didn't need any. We did add a quart of oil every twelve hours.

"Our engine just ran perfect, and everyone was impressed. We were at Oshkosh the first four days and estimate that 50,000 people inspected our installation while we answered the same questions over and over. We're still getting mail from some of them."

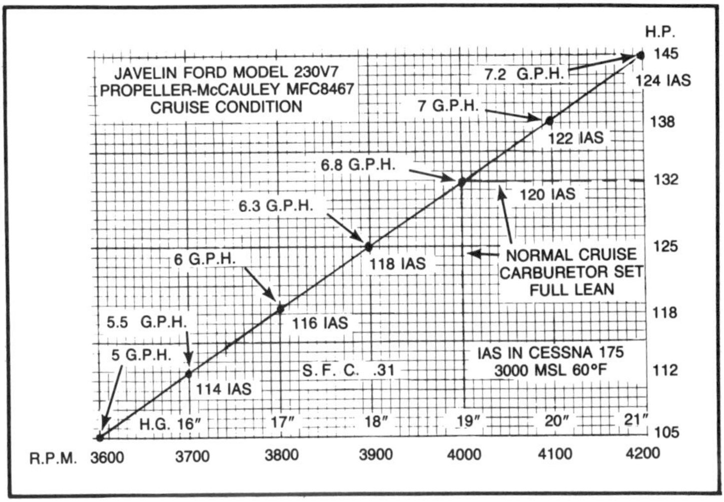

Fig. 2-14. The Javelin Ford engine of 230 cu/in replaced a geared GO-300 Continental of 300 cu/in, original equipment in the Cessna 175. With the same propeller, initial climb rate went from 500 fpm to 1500 fpm at full gross load, and operating costs dropped sharply. (courtesy Javelin Aircraft)

"Any late bulletins on the V-6?"

"Well, as stated on page 78 of our Handbook, we needed colder spark plugs. Ford now has them—Motorcraft AGSP32. This is the flat seat plug for the early 230-V6. For the 1983 model and onwards

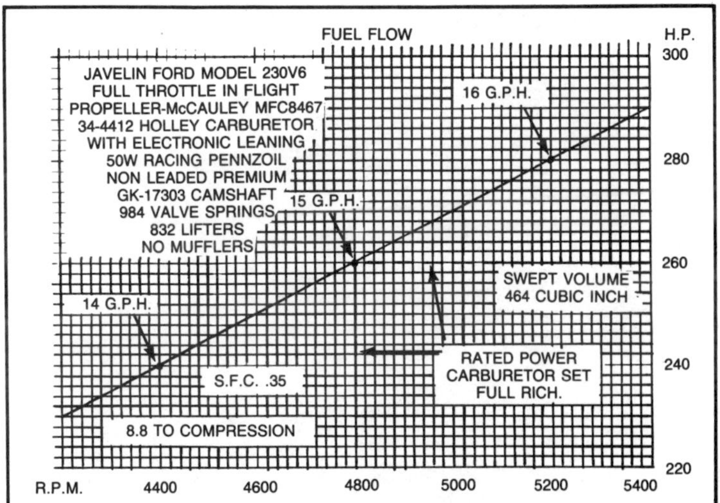

Fig. 2-15. Fuel flow. By way of comparison, the Continental 0-470 flows 27 gph (sfc of .7) at 230 hp and weighs 488 pounds.

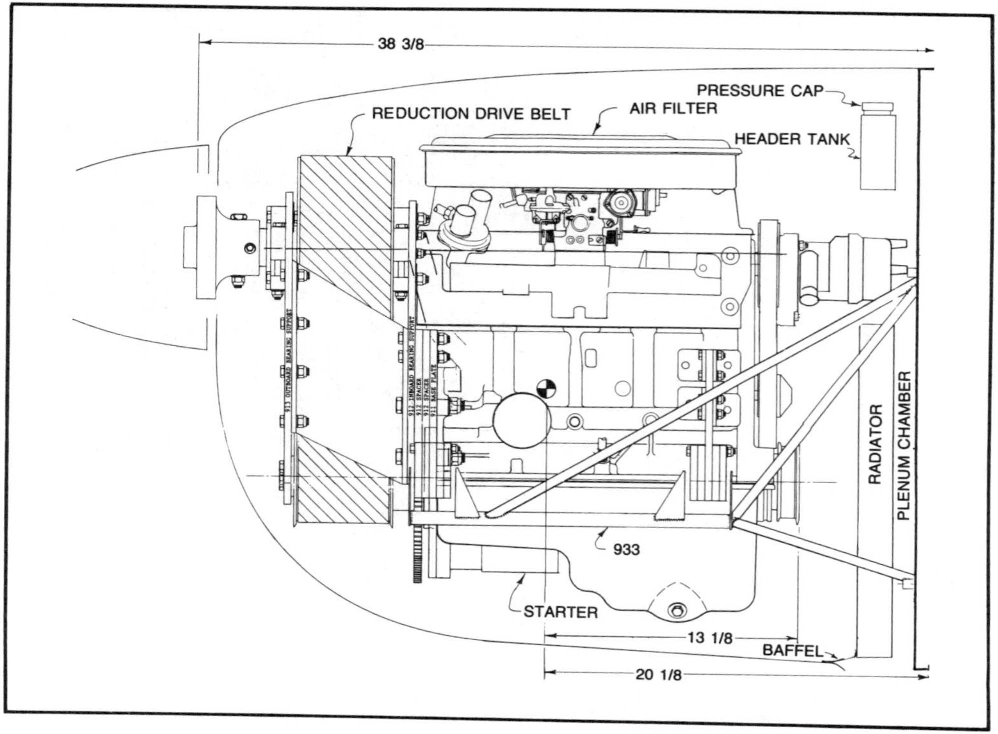

Fig. 2-16. Scale installation drawing shows final radiator position. The Javelin Ford Handbook also shows pusher installations.

Fig. 2-17. The Javelin Ford V-6 installed in a Wag-Aero 2 × 2 Sportsman. This four-placer is a homebuilt copy of the popular—but now rare—Piper Family Cruiser produced during the late '40s.

the same plug is needed with a tapered seat, and the part number is AWSF32. We have flown these plugs for many hours on non-leaded and non-leaded premium, as well as LL100 av-gas, and they are perfect."

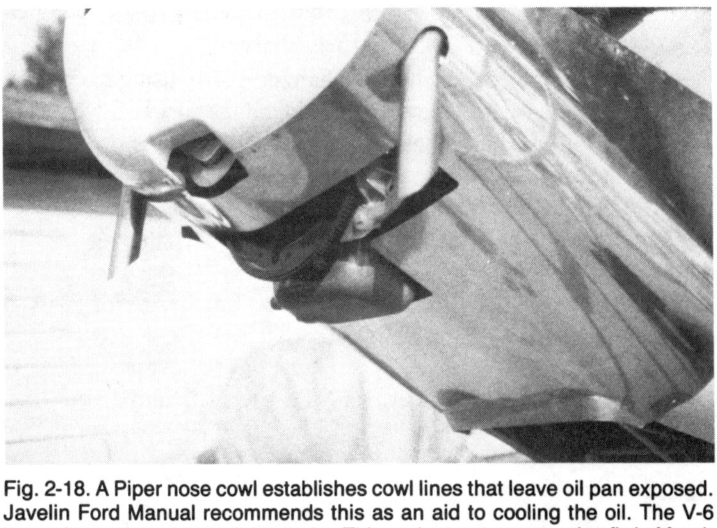

Fig. 2-18. A Piper nose cowl establishes cowl lines that leave oil pan exposed. Javelin Ford Manual recommends this as an aid to cooling the oil. The V-6 is a quiet engine with straight stacks. This project was expected to fly in March 1986, but builder asked for anonymity to limit the phone inquiries.

Fig. 2-19. The Wag-Aero Sportsman kit, less engine and propeller, is priced at $11,263.45 from: WAG-AERO, Inc., 1216 North Rd; Box 181, Lyons, WI 53148. Phone: (414) 763-9596. The original Piper PA-14 Family Cruiser was underpowered with the 0-235 Lycoming of 108 hp.

"You haven't changed your mind about turbocharging, have you?"

He shook his head. "We flew turbochargers for five years before concluding that it was all a mistake. You must double the size of the radiator and double the fuel flow to cool it. Unless you live at 8,000 feet, you don't want a turbocharger.

"Speaking of heat, the heat exchanger we are using to cool the oil is a great success. It now seems ridiculous that we ever tried an oil radiator on a liquid-cooled engine. Also, there is no need for the carburetor heat source as shown in our Handbook. We have flown in rain at low temperature and cannot get carburetor ice. Without the insulating shim, the carburetor is running at 145 degrees F."

"Can you obtain a Supplemental Type Certificate for the V-6 installation in the Cessna 175?"

Dave shook his head. "No. You must have an Approved Type Certificate before you can get an STC to put it into an ATCed airplane."

Costs

Javelin Ford conversion costs cannot be firmly established. Almost without exception these builders will obtain their engines from a salvage yard. It isn't difficult to find good Ford engines taken from wrecked cars and trucks because millions have been built and both versions are currently in production. Prices seem to range from about $600 to $1,200. A brand new one from a Ford dealer could go as high as $2,500.

However, one does not build up these engines piece by piece. There is surprisingly little "converting" to do. You do not even take the heads off of the V-6. You do remove the head from the Model 116, but only to open plugged head-gasket holes for improved cooling. Try to get a low-mileage engine from a car totalled by rear-end damage, and run a compression check if the engine has already been removed from the car.

Before you do anything, of course, obtain the Javelin Ford conversion manual. If you can manage a trip to Wichita and a visit with the Javelin Aircraft people, that will be time well spent. The Javelin Ford manual is priced at $56 and contains all scale drawings for the conversion of both the Escort engine and the 230 V-6, along with a running account of all the development experiences encountered while designing and testing these conversions. The Javelin Ford information package is $5.00.

The reduction drive kits are priced at $2031.00 for the Model 116 engine, and $1,872.00 for the V-6 engine. A choice of three reduction ratios are available for the V-6, depending upon application. The Model 116 Escort engine reduction drive is available only with a 2:1 ratio.

For additional information, contact:

Javelin Aircraft Company, Inc.
Municipal Airport,
Augusta, Kansas 67010
Phone: (316) 733-1011

The following mods are necessary to convert the V-6: 1) Replace the valve lifters with non-pump-up lifters; 2) Replace valve springs; 3) Replace camshaft; 4) Delete EGR system and open intake manifold; 5) Replace carburetor with a Holley 34-4412; 6) Modify accessory pulleys to slow down the alternator; 7) Install the propeller reduction drive, and 8) Install the inboard engine mounts.

Under Construction

As of this writing, manuals and/or reduction drives had been purchased for the following according to Javelin Aircraft:

Aircraft	Number
Acro Sport	2
Acroduster	2
Bede 4	7
Bakeing Duece	2
Bushby Mustang II	4
Wag-Aero Sportsman	30
Coot	3
Cozy	3
Defiant	14
Dyke Delta	2
Durand Mk4	2
Falco	2
Glassair	6
Scale Hurricane	3
LongEze	4
Morrisey	4
Ospery II	3
Scale P-51	6
Pitts	2
Polliwagon	3
Pietenpol	4
RV-4	4
Skybolt	3
Starduster II	3
Scale SE-5	8
T-18	12
Tailwind	4
Volmer	2
Zenith	10

Plus other one-of-a-kind.

Chapter 3

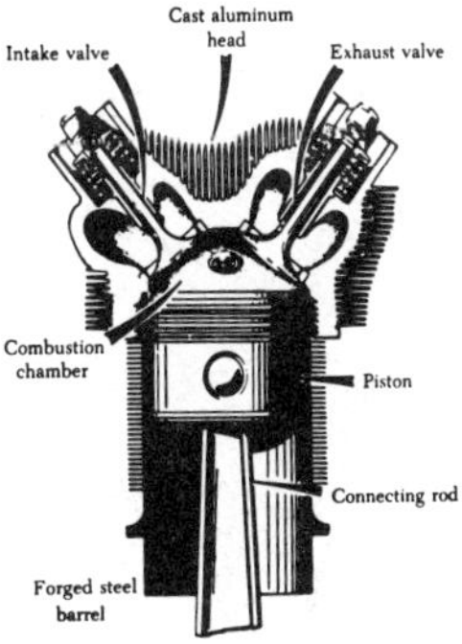

The HAPI Engines

The successful engine people seem to have one thing in common: Their creations work because they have tried all the things that *don't* work. The leaders in converting automobile engines to aircraft use have spent years developing the powerplants they market and they are quick to tell you about the engines they have trashed due to "good ideas that didn't work out." They will also tell you that they have greatly benefitted from the experiences of others. Rex Taylor says, "We at HAPI do not intend to imply that all the changes necessary to convert the VW engine to aircraft use are products of our own ingenuity . . . What we have done is put together our particular combination of conversion ideas which have been proven."

HAPI Types

HAPI engines are available in eight versions of the 1680cc and 1835cc piston displacement sizes (Fig. 3-1). "The eighteen thirty-five cubic centimeter version is the largest practical displacement for a reliable VW conversion," Rex says.

The *HAPI E Series* engines are for airplanes without an electrical system, without starter or generator. Weight is 145 pounds. Model 50-E is a 1680cc conversion which is rated at 55 hp at 3200 rpm. Model 60-E is the 1835cc engine rated at 60 hp at 3200 rpm (Fig. 3-2).

Standard equipment includes oil pump, oil cooler, carburetor

Fig. 3-1. There are eight models of HAPI Volkswagen conversions available at 55 or 60 hp, single or dual ignition. A 75-hp turbocharged model is offered for those who insist on it. (courtesy HAPI Engines, Inc.)

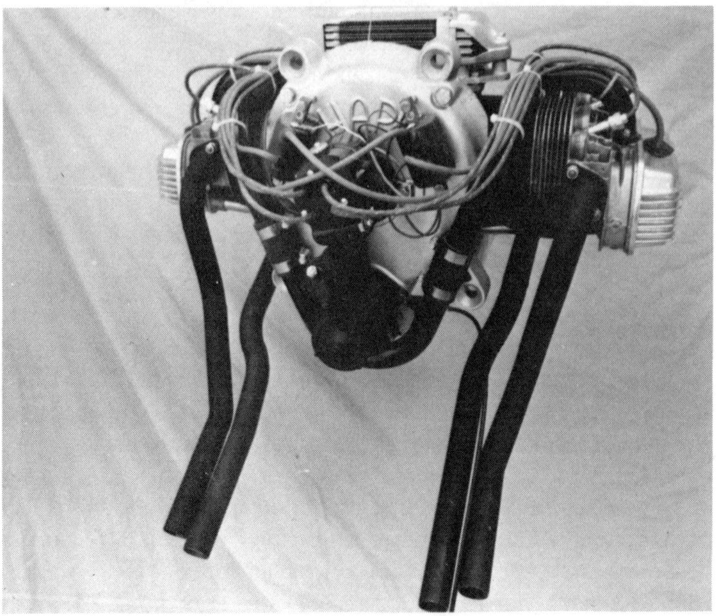

Fig. 3-2. Rear view of HAPI VW 1835cc conversion; largest practical displacement for a VW, HAPI's Rex Taylor says. (courtesy HAPI Engines, Inc.)

(with mixture control), intake manifold, lower cylinder shrouds, single magneto, and magnesium accessory castings. Options are shielded ignition harness and spark plugs, dual ignition, float-type carburetor (in place of the Posa), tuned exhaust, and 10-Amp alternator.

The HAPI Dash-2 Series engines are the best sellers, incorporating as standard features the options most builders want. These are the engines to choose if you are planning to install a radio and aircraft accessories requiring an electrical system. The Model 50-2 is 1680cc, 55 hp, and single ignition, while the Model 50-2M is the same engine with dual ignition. Similarly, Model 60-2, the 1835cc engine of 60 hp has single ignition, and the 60-2DM 1835cc has dual ignition.

Standard equipment in all Dash-2 engines includes all features listed for the E Series standard-equipped engines, plus vibration-dampening mounting, 20-Amp alternator with harness and regulator, electric starter, and shielded ignition. Options are a float-type carburetor, tuned exhaust for some models, a windage tray for aerobatics, and two-stage oil pump with full flow filter system. The Dash-2 Series engines weigh 158 pounds.

The HAPI models 60-S and 60-SD, single and dual ignition, respectively, are 60-hp VWs of 1835cc designed specifically for the Corby Starlet, an all-wood aerobatic aircraft which HAPI plans to market in kit form. These engines are similar to the Dash-2 models, but have no starter or starter ring. They feature a lightened flywheel, and a windage tray in the crankcase to hold oil in the sump during aerobatics.

Actually, HAPI developed the turbocharged models 60-TC-2 and 60-TC-2DM, but it seemed apparent (at least to me) that Rex Taylor was not satisfied with the reliability factor of turbocharged VWs, including his own. The HAPI VW-TCs incorporate automatic wastegate action to control potentially dangerous boost pressures and keep them within acceptable limits. The turbo is supplied with oil from a special oil pump that also scavenges the oil from the turbo and returns it to the sump. This eliminates the troublesome problem of blowing turbo seals and pumping oil overboard, a problem common in some other turbo VW systems. The HAPI turbo develops 75 hp at 35 inches of manifold pressure (Hg), which is the maximum allowable for takeoff.

HAPI Engine Kits and Sub-Kits

Although a ready-to-fly HAPI is the best guarantee of a sound engine, professionally assembled, you may build up your own from a complete kit or, starting with a good used VW engine, purchase the conversion parts as a sub-kit from HAPI to create your powerplant.

The Longblock Sub-kit No. 1 contains everything necessary to build an 1835cc engine, including a brand new crankcase, matched and balanced forged steel crankshaft, the HAPI C-60 camshaft, balanced piston and rod assemblies, cylinders, completely machined and assembled cylinder heads, propeller hub and seal, lower cylinder shrouds, plus all other parts, nuts, bolts, washers, gaskets, etc.

The Conversion Sub-kit provides the special aircraft parts needed to convert your engine to aircraft use: balanced piston and rod assemblies, forged crankshaft, aircraft cam—everything you need to turn your VW engine into an aircraft longblock. Naturally, you will have to have some machine shop work performed on your automobile case and heads. Best bet is have HAPI do this for you; a local machinist will not be familiar with the requirements.

The HAPI Sub-kit No. 2 contains the carburetor and oil cooler with attendant hardware. This group of items is common to both the "E" and Dash-2 Series engines, and is the next group to be added to your longblock.

The HAPI Sub-kit No. 3 completes your engine either as an

Fig. 3-3. Rex and Phyllis Taylor in their HAPI-powered Dragonfly demonstrator. The Taylors market Dragonfly kits for $12,995, which includes engine, prop, instruments, paint—the works. (courtesy HAPI Engines, Inc.)

Fig. 3-4. Phyllis Taylor runs the office at HAPI headquarters on the Eloy (Arizona) Municipal Airport. (courtesy HAPI Engines, Inc.)

"E" or a Dash-2. Sub-kit No. 3 for the "E" includes a Slick 4216 magneto, unshielded harness, mag drive casting and attendant hardware, spark plugs, plus the nuts and washers to get the E engine ready for startup. Sub-kit No. 3 for the Dash-2 engine, available for both single and dual ignition, contains the accessory case, the flywheel with alternator rotor installed, alternator stator and regulator, starter, magneto, the shielded ignition harness and spark plugs, plus all the rest of the hardware required. If dual ignition, one also receives all the parts of the electronic ignition system, the secondary ignition system being totally independent of the magneto system (Figs. 3-3 through 3-6).

HAPI Engine Prices

Early in 1986 HAPI engines were priced at the factory as follows:

Model 50-E	$2450.00
Model 60-E	2550.00
Model 50-2	3050.00
Model 50-2M	3195.00
Model 60-2	3150.00
Model 60-2DM	3550.00
Model 60-S	2995.00
Model 60-SD	3395.00

Kits for the above are priced $300 to $500 less.

Fig. 3-5. Son Pat Taylor is responsible for HAPI engine manufacturing; Pat or Rex may be reached by phone during working hours to answer builders' questions.

Fig. 3-6. HAPI engine production; all are test run prior to shipment. One may purchase a ready-to-fly HAPI VW, or build from scratch with HAPI components. (courtesy HAPI Engines, Inc.)

Longblock kit, single ignition............$1510.00
Longblock kit, dual ignition..............1620.00
Conversion Sub-kit........................675.00
Sub-kit No. 2 (carburetor and oil cooler).....295.00
Accessory Sub-kit No. 3, E-engine.........465.00
Accessory Sub-kit No. 3, Dash-2 engine.....965.00

HAPI Engine Components

Engine cases: $289.00 and $319.00.

These are 1600cc factory new crankcases with 8mm case savers installed as standard equipment, shipped in original crate and preserved in cosmoline. If you plan to build your engine larger than 1600cc, HAPI can machine your case prior to shipment.

Forged crankshaft: $149.50.

Rex Taylor says that cast cranks are dangerous. The HAPI crankshafts are forged steel, magnafluxed, reground to close tolerances, and electronically balanced. HAPI will taper the front of the crank free if you order your prob hub at the same time. These are brand new, German-made components.

Propeller hub: $125.50.

The HAPI parts catalog points out that the prop hub is one of the most important items in your airplane, transmitting the engine's output to the propeller for the thrust to produce flight. This assembly is precision machined from 8620 steel, and includes locking bolt, washer, and matching aluminum front plate. Hub bore is tapered three degrees to mate with HAPI crankshafts. The steel parts are finished in gold cadmium plate. Standard six-hole four-inch diameter bolt pattern for 5/16th-inch bolts, with two-inch center hole in spacer.

Connecting rods (set of four): $74.95.

These rods were originally engineered to withstand the stresses of all-out stroked racing engines encountering high loads at high rpm. These are forged rods, balanced for matched pin-end weight, and then balanced for total weight, with variations maintained within 1/4 gram. Pin and crank ends are kept within a tolerance of 0.0003.

HAPI C-60 camshaft: $69.50

Since cams ground for automobile use work best at much higher rpm than those in aircraft engines, HAPI regrinds brand new cams to a computer-designed contour that produces maximum torque in the 3,200 to 3,600 rpm range where, ideally, peak horsepower and optimum propeller efficiency coincide. Cams come in "plus" and "minus" sizes to compensate for timing gear lash. No exchange is required.

HAPI Cylinder heads, complete:$249.50 per pair to $499.50.

Bare cylinder heads are available for $199.50 per pair, but a better buy is heads with new valve springs, intake valves, stainless steel exhaust valves, keepers and retainers, all assembled with the valves lapped and ready to use at $249.50 pair (Figs. 3-7, 3-8).

To save building time, HAPI offers heads machined for 92 mm cylinders and 8.5:1 compression, single or dual port, $299.50; modified for dual ignition, with special rocker covers, ignition harness, and spark plugs, $499.50 per pair (dual port).

The components listed above represent a very small part of the items offered in the HAPI catalog, but they should give you a good indication of overall costs. In most cases it is probably best to purchase a complete engine, because Rex and son Pat, along with their shop people, have so much experience with these conversions that one is certain to get a problem-free powerplant.

However, many homebuilt airplane constructors proceed on a pay-as-you-go basis, spreading the labor and cost over a period of a year or so (the day of the five-year homebuilt airplane project seems to be passing), and that is why HAPI offers all engine components in a separate price list.

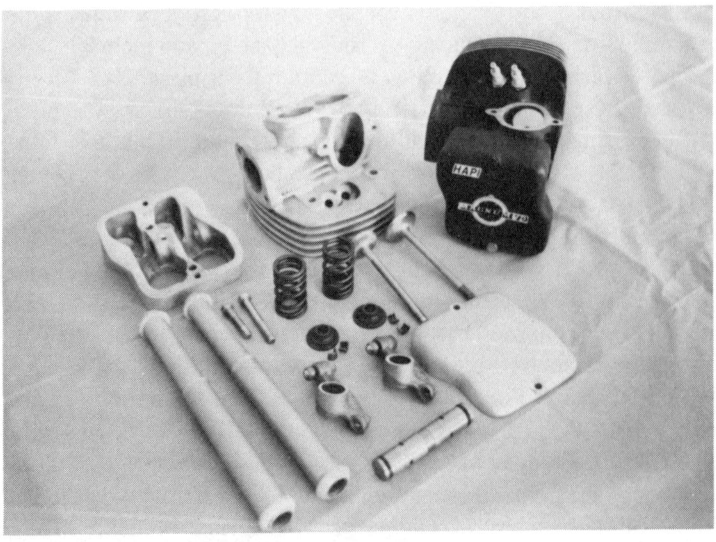

Fig. 3-7. HAPI cylinder head and valve assembly; valve lifters are hydraulic. (courtesy HAPI Engines, Inc.)

Fig. 3-8. HAPI is equipped to perform all machine operations on its engines. Rex Taylor left Lockheed to build VW racing engines; as a pilot, recognized the need for low-cost, reliable engines for homebuilts and HAPI development began in the '70s. (courtesy HAPI Engines, Inc.)

Engine Accessories

If you add an electrical system with alternator and starter, you'll begin with the accessory case. It is cast of aircraft magnesium, and serves as a means to attach the magneto, alternator, and starter. It also provides engine-mount attachment points. This case will mount either Slick or Bendix single or dual magnetos. The engine mount pattern is standard A-65 through 0-200 Continental, and it uses conical rubber bushings for vibration isolation. This case alone is priced at $156.95. The 20-Amp alternator with regulator is $139.50, and the aluminum flywheel with starter gear is $93.95, while the electric starter is $98.00. The alternator, regulator, and flywheel are contained inside the accessory case.

Surefire Ignition System

According to Pat Taylor, airplane ignition systems have not changed since before WWI. "HAPI has developed a modern electronic ignition system and proven it in over 5,000 hours of absolutely reliable and trouble-free service over the past three years as the secondary system on our Dash-2 series engines.

"We have now designed an electronic interface allowing the electronics to run directly off our solid state alternator. Now we can throw away the old-fashioned, expensive, and troublesome magneto.

"This represents a major breakthrough in aircraft ignition system, and has the following features:

- ☐ Right or left rotation, with any lag angle for starting.
- ☐ Hotter spark; fires .030 plug gap vs. .016 with magneto.
- ☐ There are no points, rotors, or other moving parts to wear out or get out of adjustment.
- ☐ Generates no electrical noise; shielded harness and plugs are not necessary.
- ☐ Runs totally independent of battery after starting.
- ☐ Provides easier starts; 48,000 volts fires cold engines instantly.
- ☐ Better fuel economy. Hotter spark and wider gaps mean better fuel burn.
- ☐ Fits in place of the Slick or Bendix mag and uses the same mechanical drive.
- ☐ Single or dual system available.

"This system, single ignition, is priced at $395.00."

Carburetors

HAPI offers a float-type carburetor and two versions of POSA fuel injector.

The POSA fuel injector carburetor is a very simple device. Since it has no float or float bowl, it cannot "flood" in the usual sense. Without a venturi or butterfly, icing is unlikely although HAPI recommends the installation of a carburetor heat box and muff (available for HAPI). The POSA functions well in any position, has but one moving part, and is effective with gravity feed (Figs. 3-9, 3-10). It requires no critical adjustments, and is priced at $62.95. With controllable mixture and an idle mixture air bleed, the POSA becomes the HAPI SuperCarb at $115.50.

For installations where gravity fuel feed is not used and a fuel pump is employed, HAPI offers a little float bowl carb.

The HAPI parts catalog is unusually well illustrated and contains dozens of items for a VW conversion that I do not mention here.

I should also mention that Rex Taylor has written a book, *How to Build a Reliable VW Aero Engine*, which is available from Rex

Fig. 3-9. Posa carburetor is simple, reliable; manufactured by HAPI.

for $13, and which has been very successful because it details exactly what its title proclaims.

Preliminary Decisions

The most critical time for the homebuilt airplane constructor

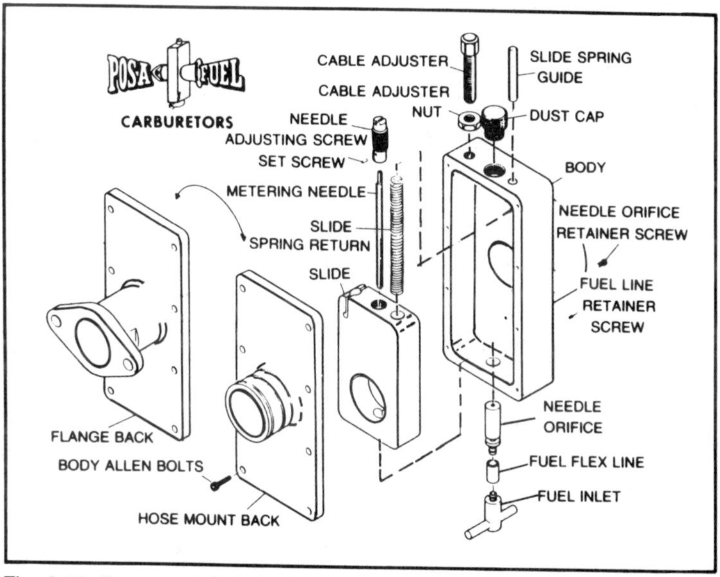

Fig. 3-10. Posa exploded view.

may come before any labor or money is invested. Proper decisions concerning the airplane and its powerplant—"proper" meaning a design and materials commensurate with your skills (both building and flying skills), and a dependable engine of sufficient power—mark the difference between success and varying degrees of failure. It should be noted that there are too many "semi-failures" flying around out there or tied down somewhere with "For Sale" signs on them. I have been reporting on homebuilt airplanes for more than 20 years; I became a member of the EAA (#11,645) when Paul Poberezny was writing to us in longhand, and I've observed firsthand many of the more serious mistakes in judgment of which first-time homebuilt airplane constructors are guilty.

First of all, don't select a design that "tenses you up" to fly it. It will spend most of its time in the hangar eating up the fixed costs of ownership and seeming to challenge, rather than invite, you to fly. For most of us, the high-performance machines quickly lose their appeal. We take some satisfaction from the fact that we have mastered the less forgiving aircraft (i.e., flown them for a few months without killing ourselves), but then we begin to find a reason or two to sell or trade for something that is less demanding and more fun. The gentler machines are also easier and less expensive to maintain, and will safely operate from fields that the slippery designs must pass up.

Keep in mind that every airplane is a bundle of compromises. You get something by giving up something. If you want an airplane that will smoke along at Warp-9 speed, be prepared to sacrifice the relaxed pleasures of low and slow.

Another important consideration is the power requirement of the design you choose. First, we determine what is available in the proper power range, then we look at initial cost, the dependability factor as well as that may be determined, and future maintenance demands.

"Dependability" may be enhanced by good maintenance and sound operating procedures. Generally speaking, the simpler a piece of machinery, the more dependable it is and the less costly it is to operate (as the late Bill Lear once put it, "You'll never have to maintain, inspect, or replace anything you leave out"). But as mentioned, you must compromise. You can get a lot more efficiency from a liquid-cooled engine than from an air-cooled powerplant. The liquid-cooled engine will run cooler and give significantly better gas milage, but its cooling system and propeller reduction drive add cost, weight, and something else to maintain. The laws of physics are the same for everybody, and no one gets a free ride. If you plan to supercharge your VW conversions, be ready to pay the price.

Will you buy a ready-to-fly engine, professionally assembled by a conversion expert, or will you opt to save the difference, start with a bare crankcase, and build up your own—perhaps spreading the cost painlessly over a period of a year or so?

Building a VW Aero Engine

If you expect to build your own engine you will need a clean, dust-free place in which to work, proper tools, reference material to guide you through each operation, the services of a reliable machinist, and the patience to do the job right. As Rex Taylor puts it:

"If you have the above-mentioned prerequisities, you can build your own engine. On a cost-saving basis, if you place a reasonable price per hour on your time, it will probably cost you more to build your own engine than to buy it. In terms of pride and accomplishment, you will gain an understanding of your engine, and indeed may find that working on engines is a very enjoyable pastime.

"If, however, you are an impatient person who does not enjoy working to exacting tolerances, or do not like to get your hands oily, or perhaps are one of those with a 'that's good enough' attitude who is satisfied with less than perfect in the jobs that you do accomplish, do yourself a favor and buy your engine. There is only one way to build an engine right, and that is to build it the very best it can be.

"To elaborate on the five essentials: Number one, the clean, dust-free area need not be large. A corner of your garage should be sufficient. I assume that you have some kind of sturdy table at a comfortable working height, and a bench vise. Two very useful tools in the building of a Volkswagen engine are an engine buildup stand, which holds the engine in any position, and a drip pan such as used under automobiles to catch dripping oil on garage floors. The pan is used on top of the bench as a spotlessly clean surface for building up subassemblies and laying out your tools. Your tools should be cleaned in solvent, dust and dirt-free, before using them on your engine . . .

"Since this process will take time, your engine and its parts will require protection . . . We have found that plastic garbage bags make excellent dust covers, capable of enveloping a complete engine on its assembly fixture, and really protecting it from dirt until the next time you can work on it.

"The tools required will represent an investment of perhaps

one hundred and fifty dollars. You will need a three-eighths drive socket set, and a set of box/open-end wrenches in metric from 10 millimeter to 17 millimeter sizes. You will also need a ring compressor, screwdrivers, pliers, Allen wrenches, and, most important of all, an accurate torque wrench. If you think you can just guess your way through and avoid buying a torque wrench, forget building an engine at this point; you don't have the right attitude, and you are going to produce an undependable piece of junk. Condition yourself to do the job totally right, or don't do it at all. You will require a few more tools such as a propane touch, possibly a gear puller if you have to pull your propeller hub for some reason after assembly, a dial indicator, a two to three-inch micrometer, and an ohm meter for timing your magneto.

"Liberate some of your wife's baking pans to store nuts and bolts as they are removed from the engine. Separate them into engine sections, such as head bolts, crankcase bolts, internal engine parts, etc."

The reference materials needed should be satisfied by acquisition of Taylor's book, a step-by-step manual that will allow anyone (with the proper attitude toward this kind of work) how to assemble a top-quality VW engine conversion. There are other publications available that may be helpful, such as the Type II Engine Maintenance Manual, available at your Volkswagen dealer. If you have had little experience in engine work, you should start by looking in the public library for a good book on four-cycle engine theory, or reference TAB book 2427, *Maintaining and Overhauling Lycoming Engines—2nd Edition*, since basically all four-cycle engines operate on the same principle.

Taylor continues: "The machine shop services or machining ability and availability you will find necessary are probably the largest stumbling block for some of you who live in areas of the country where Volkswagens are not quite so common as they are in California. There is quite a bit of modification and machine work necessary to build anything larger than sixteen hundred and eighty cubic centimeters (1680cc). With the larger bore sizes, the crankcase and cylinder heads must be bored out to accept the larger assemblies. While the necessary operations are not too complex or difficult to do, to have them done in a shop that has no previous experience with these operations usually proves very costly, because the shop has not developed holding fixtures and cutting tools to do it in an efficient manner as in a production run. Usually, the shop will charge you by the hour, and that will be expensive.

"There are many firms that do this work on a routine, daily basis, at very reasonable prices, and turn out good quality work. Some of you may have to send off your parts for machining. HAPI offers these operations at competitive prices. Should you be fortunate enough to have your own lathe and milling machine, or access to such, you may well be able to perform these jobs yourself. There is one point to consider here: If you louse it up, you throw it away at your own expense. If a responsible shop louses it up, they must replace it at no cost to you.

"The last requirement is the most important of all. Those of you with patience, dedication to the task at hand and the will to finish, can probably get by very well without having the most ideal conditions to work in, the best tools to work with, an abundance of reference materials for every question, the availability of an expert machinist, or machine tools of your own. If you are not willing to spend the time necessary, or don't have a good track record for finishing projects you start, you probably won't finish this one either. If you don't finish and decide to sell out, you'll undoubtedly take a considerable loss on your investment. So be honest with yourself about yourself before you start."

You still have some decisions to make. You may start building an engine with all new parts, or you may use a certain amount of used parts in the process. There are advantages and disadvantages each way. New parts are more expensive, but are easier to find and you don't have to clean them of accumulated grease and dirt. Used parts are sometimes found at low cost if you are a good scrounger, although you never really know what you have until you've spent hours in teardown, cleaning, and inspecting for damage that will render the part impossible to rebuild to aircraft standards. Sometimes it is possible to find a used engine that is very good inside; others turn out to be junk, in which case you have wasted your money because most used parts are sold on an "as is" basis.

If you plan to work with a used engine, don't hurry your purchase. Check around the automobile wrecking yards in your area and try to find a VW that was wrecked rather than one that is run out. Look for one that was totaled for front-end damage, with the engine and transaxle section undamaged. The transaxle is not important except as an indicator; if it is damaged, there is a good chance that the engine case had a severe shock and is probably damaged as well. A cracked engine case may not be found until after the engine is disassembled and cleaned.

According to Taylor, your best bet is an engine that can be started and run in the car. The engines found in the VW Squareback sedans and station wagons (which Volkswagen calls "suitcase engines") are very good for aircraft conversions. These were all built in 1500cc or larger, and most have the desired large oil galleries in the case. Best of all, they are not as eagerly sought after as the vertical-blower Bug engine and usually may be bought for less. Some of the later Squarebacks had electronic fuel injection and dual port heads. The fuel injection is not useful to the aircraft engine converter, but the rest of the engine is very suitable. It is especially good if you want to build a turbocharged engine (many VW engine converters advise against turbocharging. If you are uncertain about the use of such devices and how they work, I suggest that you read TAB book 2306, *Aircraft Turbocharging*, by Paul Garrison).

Rex Taylor can provide you with a list of engine identity numbers, including those that are most desirable for aircraft use. He also offers some of the means of differentiating between the various cases and other parts you will need. New parts are much easier to obtain, but you must know exactly what you want. There are all too many suppliers of replacement parts for Volkswagen now who are turning out parts that look like the original, but are real junk, Taylor says. "In order of quality, the best parts are usually German," he told me. "The next best come from Japan, and the real junk is from Brazil. This is not a hard and fast rule, however; each of these countries also turns out some parts of good quality. The super bargain Volkswagen parts advertised in some catalogs are usually of very shoddy quality, even for automobile use, and are poison in an airplane. Price alone is not always a good indicator, but is a fairly reliable one.

"At HAPI, we have been offered a set of gaskets for as little as 93 cents wholesale, but we still use the gaskets that cost five times as much, simply because we want the oil on the inside of the engine. The Volkswagen gasket set consists largely of rubber seals, and you must have high-quality, live, heat resistant, long-life rubber, instead of the hard and inflexible, dry junk in the cheap sets offered. If the parts you obtain will not do the job properly, they are not cheap even if you get them for nothing."

Rebuilt Components

There is another source of parts: the rebuilt cases, reground crankshafts, rebuilt cylinder heads, and even rebuilt connecting rods.

The rebuilt cases can be a good alternative to a new case if they have been aligned-bored and reconditioned by a reputable firm adhering to high standards, according to Taylor. A reconditioned case may serve as well as a new one. The key is to know with whom you are dealing. One top company is Rimco, whose cases are equal to new ones (and the cost is nearly so).

Cylinder heads are routinely rebuilt for automobile use. They are rebuilt with salvage methods which include helicoil inserts in spark plug holes and cracks repaired by heliarc welding, both practices spelling trouble in aircraft use. You may find a pair of heads on a used engine with no such fixes, but be aware that the used engine may have other less-than-airworthy things done to it in the past. New cylinder heads are not too much more in price than top quality rebuilds, and with new ones you may be certain of what you have.

Crankshafts are a different matter, with the used ones being the best ones within certain limits. For many years, Volkswagen built their engines with a high quality forged steel crankshaft with bearing journals surface-hardened. These shafts are readily available, and most will have serviceable, standard size bearings. Such a shaft, after magnafluxing, is reground to .010 undersize on the rod and main bearings, and will be far superior to the soft, cast steel cranks found in later model VW engines. A cast crank, subjected to bending or shock loads, is brittle and will break like a piece of glass (Taylor's phrasing), whereas a forged crank will bend without breaking. Clearly, a broken crankshaft can result in loss of your propeller in flight. The 1500cc crankshaft, VW part 311-353-A, is recommended by Rex Taylor because, he says, it has cross-drilled oil-feed holes to the bearings. It does not have the elongated slots as did later cranks. The bearing loads imposed by aircraft service cause the elongated-slot crankshafts to wear out-of-round due to the smaller amount of bearing surface contact.

Rebuilt connecting rods are cheap, but should be avoided. There are too many things that can be wrong with them that may not be apparent on inspection. HAPI uses a very expensive con rod, compared to the rebuilts, but Taylor claims that no HAPI engine has ever suffered a connecting rod problem of any kind and therefore the extra cost is justified.

Most of the remaining parts in the engine you build will be new except for those stock parts that are used in non-critical areas.

The other parts that are unique to the Volkswagen conversion are the propeller hub and the various castings or machined plates necessary to cover, or alter the function of, various openings in the Volkswagen crankcase.

At this point, you will have to decide what kind of conversion you are going to build; that is, whether or not you want dual ignition, an electrical system, etc. Most homebuilt airplane designers would probably be happy if the builder would stick to the simple versions of their aircraft, which is to say, the lighter-weight versions which perform better. Keep in mind that while a starter, alternator, and accessory case are nice, this requires the use of a voltage regulator, battery, battery cables, ammeter, master switch, master relay, and perhaps a few more goodies. The total weight added to your airplane will probably exceed 25 pounds. (A turbocharger will add another 30 pounds, including its complex plumbing.) All of this weight is normally concentrated in the nose of the aircraft, and many times requires moving the engine forward to make room for everything. Then, one not only has a lot more weight up front than the designer ever planned, but it is farther forward to complicate the center of gravity problem even more. None of these accessories add any horsepower, and your airplane will know it at takeoff.

Tuned Exhaust

There is an area where you can get a solid horsepower increase without harming the engine and without going to any great expense. This is known as *exhaust tuning*. To understand what exhaust tuning is, try to visualize what goes on inside the engine's combustion chamber. Let's start with the intake stroke. The intake valve is open, and the piston is on the downstroke, creating a vacuum within the cylinder. In addition, the atmospheric pressure is pushing air through the carburetor, mixing it with fuel and forcing it through the intake manifold into the vacuum in the cylinder. Ideally, this would charge the cylinder at the bottom of the stroke with a cylinder full of fuel/air mixture ready for the compression stroke, which would be at full atmospheric pressure. In practice, however, this ideal condition is impossible to achieve, because the intake valve closes slightly before the piston reaches bottom dead center (BDC) and the cylinder has not reached a pressure equal to atmospheric pressure. So the piston is ready to start the compression stroke but must move up the cylinder a certain distance before atmospheric pressure can be reached and actual compression of the fuel/air mixture can start.

Clearly, anything that will result in a better vacuum in the cylinder, or more atmospheric pressure at the bottom of the intake stroke, will result in more fuel/air compressed and fired. The more fuel/air charge we can cram into the cylinder, the more power we can develop with each firing.

There are many ways to increase the size of this charge. Most of us are familiar with the old single-barrel carburetors on our automobiles; later came two barrels, and now modern cars have four-barrel carburetors with progressive linkages. The main purpose of this is to really open up the intake manifold to allow atmospheric pressure to charge the cylinders quickly, and with the largest amount of fuel/air mixture possible.

It is more difficult to charge the cylinder efficiently at high rpm than at lower speeds, because air is a fluid and tends to resist any change in its condition, either in motion or at rest. Since we are alternately starting and stopping this movement of fuel/air into the cylinder with the intake valve, the mixture must again move from a still condition.

Tuned exhaust systems are based on the principle that a fluid in motion tends to stay in motion. If you should send a quick, hard blast of air into one end of a long exhaust pipe, then instantly cap the end, the air rushing through the pipe would, due to its tendency to remain in motion (inertia), actually create a partial vacuum in the capped end of the pipe. The turned exhaust system uses this principle in that the outgoing high-velocity gases will pull a vacuum in the cylinder that will help to scavenge the burned gases and get maximum effectiveness from the fresh charge of fuel/air mixture. There is a certain period of time at the transition point between the exhaust stroke and the intake stroke when both intake and exhaust valves are in the open position, which allows this scavenging action to take place. This is called *valve overlap*. The length and diameter of the exhaust pipe, the amount and radii of the bends, interior surface roughness, and other factors are all involved in the tuning process, which is really effective in a very narrow rpm range. Since engines for aircraft use have a narrow power range, exhaust tuning is somewhat easier for us.

The theoretically perfect exhaust system on an 1835cc VW would be a straight pipe with no bends, 1 1/2-inches in diameter and 84 inches in length for each cylinder. That would be known as a "full wave" system. HAPI's system is built on the half-wave length of 42 inches, which is less efficient, but Taylor believes it to be the best compromise between weight and efficiency.

"There is no way the tuned system can harm your engine," Taylor says. "It merely lets the normally aspirated engine to work much closer to the theoretical 100 percent efficiency that of course is never achieved.

"Tests indicate that the tuned system provides from 5 to 12 percent more available horsepower at the propeller. The tuned system is tucked around the engine to take up very little space, but does weigh an extra eight pounds over an engine with short, straight exhaust stacks. You have to decide whether or not the extra power is worth the extra weight.

New or Used?

If you choose to build up your VW conversion starting with a used engine, the first thing to do is get it as clean as possible on the outside, and drain all of the oil from the inside. Take it to a coin-op car wash and use the high-pressure hot water and soap. It won't hurt to get soap and water inside the engine if you are going to disassemble it within the next 48 hours. Otherwise, try to keep water out of the interior to avoid rust.

When the engine is really clean, take it home and attach it to the VW assembly stand (available from HAPI for $39.50), which allows the engine to be rotated for easy accessibility. The drip pan goes on the floor beneath the engine.

Begin disassembly with the rocker arm covers, oil sump cover, and screen. Take the screen apart and check for metal particles before discarding. The presence of metal bits is of course a warning that something inside the engine is shedding critical material.

The cylinder heads come off next and you'll perform this operation with care because there is a chance that they can be used. Loosen all the head bolts just a half-turn or so to relieve the pressure in a uniform manner. A stud that is larger in diameter than the others usually means that the original was replaced due to stripped threads in the case—not good news.

The heads will be thoroughly cleaned and inspected later. Meanwhile, remove the flywheel, and the pulley on the other end of the crankshaft. The HAPI manual describes this operation thus: "You will need two large 1/2-inch drive sockets (one 30mm and the other a 36mm-plus), and a 1/2-inch drive breaker bar. The flywheel is usually torqued to over 200 ft/lbs, so you will need a flywheel lock tool to hold the flywheel stationary while you loosen the gland nut. This tool sells for about $10. Remove the gland nut, and the pulley bolt from the other end of the crankshaft. Put two large screwdrivers, 180 degrees apart, behind the flywheel and gently but firmly rock the flywheel off its four pins. Save the pins and the flywheel. Remove the pulley in the same way.

"Next, gently remove the cylinders so that the piston and rod assembly doesn't strike the case as each cylinder comes off. After the cylinders are removed, take the wrist pin clips from the pistons and use a six-inch piece of 3/4-inch wooden dowel to drive the wrist pins far enough through the pistons to free them from the upper end of the connecting rods. There is potential here for damage to the rods; therefore, when driving wrist pins either in or out never allow shock from a hammer blow to be transmitted to the rod. Avoid this by placing your free hand around the piston and gently tapping the wrist pin while the shock is taken up by your hand. Never allow the piston to lay against the rod so that the hammer taps could possibly bend the rod. Correct rod alignment is essential and can be lost by one careless tap of the hammer.

"Now you have removed the pistons and have all the other goodies outside the case with only the actual case bolts holding the engine together. Remove all of the small nuts and bolts around the outside of the case, including the one that is often hidden under a layer of grime on the flywheel end of the engine about an inch below the bottom large nut in a kind of a well casting. If in doubt, count the small bolts; there should be 15 of them. Then remove the last six large bolts holding the case together.

"Now, this is critical: A lot of cases are junked at this point by people who try to pry them apart. Volkswagen uses precision machined surfaces where the two halves of the case meet, and you can destroy a case using improper tools for this operation. There is no gasket.

"Turn your engine stand so the case split line is vertical. There are certain areas on the engine (Fig. 3-11) where some tapping with a hammer and block of wood will not harm the case. Never—repeat, *never*—attempt to separate the two halves of the case by driving screwdrivers or any sort of wedge between them. Take your hammer and wooden block and rap the case at the points indicated in Fig. 3-11, going from one area to the other to effect the separation in an even manner. When the alignment dowels are free (the case will open about 3/8th of an inch), turn the engine on the stand so that the half with the main bearing through-bolts is on the bottom,

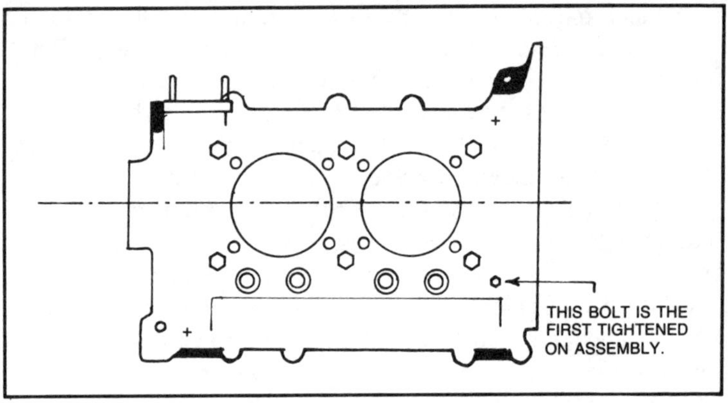

Fig. 3-11. VW case is dowled at "+" locations. The case may be bumped with a wooden block at blackened areas. Do not force it. (courtesy HAPI Engines, Inc.)

and lift off the upper half of the case. Sometimes, the crankshaft assembly tends to come with the top half, but just continue to carefully jiggle everything—using *no force*—and it all will separate.

"Lift out the crankshaft and camshaft. Look for the small dowel pins in each main bearing saddle; there should be one in each bearing saddle on the stud side of the case. Save these dowels. Set your crankshaft upright on your workbench using the flywheel as your holding fixture. Now remove the rod bolts and all the rods. Pull the old camfollowers out of the case and throw them away (the four camfollowers from the top of the case probably went on the floor anyway). Also, discard the pistons, pins, and cylinders. You will use most of the valve train if it is good, except the valves, guides, springs, and keepers, which should be replaced."

It is at this point that you determine whether you have a rebuilder or a piece of junk, and you start by measuring the amount and kind of wear on the crankshaft.

The standard size on the rods and main bearings is 2.165 inches. If there is no damage or serious out-of-round condition, and the crankshaft will regrind to no more than 2.155 inches, it will be a good, airworthy crank. If it has previously been reground in overhaul, Taylor recommends that you replace it, unless it is 2.155 inches and otherwise in perfect condition—which, he says, isn't likely. Bearing failure is usually at the #2 main or one of the connecting rod journals.

In order to measure all the journals it will be necessary to take the crankshaft to an automotive machine shop, because an arbor press is needed to remove the timing gear and the distributor drive gear.

The machine shop can also measure the rods and mains to make sure that they will clean up to no less than .010 undersize.

HAPI recommends that you have your crankshaft magnafluxed to check for hidden flaws before you spend much time and money on it. "If magnaflux is not available, there is a very simple test that can be performed to check for cracks. Suspend the crank by a piece of wire and tap it lightly with a hammer on a non-machined surface. It should ring like a bell. A cracked shaft, or a cast steel shaft, will produce only a dull 'thud' when so tapped.

Next comes a careful inspection of the crankcase, and I quote directly from the HAPI manual, *How to Build a Reliable VW Aero Engine:* "Remove the case from the engine stand and wash it so clean inside and out that you could eat off of it. If you found metal in the oil screen, you probably have trouble in the case. Check for cracks first, especially if your engine came from a wrecked car. These magnesium cases are on the brittle side, and while strong, will crack under certain stresses. Some of the things to look for are broken ears at the oil cooler mount; main bearing saddles that have suffered a bearing spin, thereby ripping out the dowel pins and enough metal to junk the case; and stripped threads around the oil screen cover plate. Also check the camshaft bearing saddles for damage and excess wear. Some of the older cases did not have cam bearing inserts, but ran the cam directly on the case material. HAPI doesn't consider this case a real good one for aircraft use unless it is new or has very low milage and wear.

"If your case is satisfactory, take it to a machine shop experienced in Volkswagen work and have the main bearings checked for size. If their condition is such that the case will clean up at .020 oversize, you are in business. If such is not the case, just take your lumps and write off that case for aircraft use.

"Any crankcase that is to be rebuilt for aircraft use should be align-bored. Volkswagen cases tend to wear in the saddle surface that is in contact with the other side of the main bearing shells. If this wear is not taken care of by align-boring, the new bearings will not be held in place with the proper amount of 'squeeze,' and consequently you will have excessive clearance in the mains even though the bearings are new. This is one of the areas where too many people attempt to cut the cost of rebuilding by omitting the boring operation, and a very short time later they are rebuilding again because the bearings didn't hold up in service. The crankshaft and bearing loads imposed in aircraft service are much more severe than those normally experienced in automotive service."

Cylinder Head Inspection

Rex Taylor makes the point that by the time one has cleaned, inspected, installed new valve guides, reground the valve seats, and accomplished the other required operations on the used cylinder heads, one usually has almost as much invested as the cost of new ones. "The used heads can't be adequately inspected until you have invested quite a bit of labor and time in them," he says. "First, the valves must be removed. Then the heads should go into a hot tank to remove grease and grime (unless you really got them clean), following which they should be glass-bead blasted to remove the carbon buildup in the combustion chambers. Then you may begin to see what you have. The two most common problems on Volkswagen heads are stripped threads in the spark plug holes, and cracks in the heads. Cracks may be most anywhere, but are usually from the exhaust valve seat to the spark plug hole, or radiating out from the valve seat in some other direction. Another common place for cracks is in the boss around the exhaust valve guide.

"Other problems you run into include bad valve seats which must be replaced, and heads that have previous cracks which have been welded.

"The camshaft is usually found to be serviceable, but the question is whether it is the proper grind to produce the torque in the power range we want. It can almost always be reground to a special aircraft grind such as HAPI's C-60 specification."

As you can see, there are many pitfalls to spoil one's hope of obtaining a good engine for little money. It does happen, but chances are that you will have to replace a good part of any used engine—and you could lose it all.

There are many sources for new parts, with your local VW dealer being both reliable and expensive. While the dealers' Volkswagen parts are genuine, there are also genuine Volkswagen cases and heads sold through non-dealer channels that are much less costly. There are also some heads from Japan that are good, Taylor says, but they are not genuine Volkswagen parts. Crankcases are made only by Volkswagen due to the great cost of tooling up for such a casting and then machining it. There are genuine cases available from Volkswagen Brazil and Mexican factories, and they are imported and sold through many non-dealer outlets.

As mentioned, HAPI can supply new cases and heads, as well as all other VW parts. One advantage in buying a new case is that it will have steel inserts which eliminate the old problem of head studs pulling out (as long as proper torque values are used). The new heads should have the exhaust valve guides replaced with bronzanium guides, then the exhaust valve reground to those guides. HAPI recommends a 1/16-inch contact surface with their stainless exhaust valve.

"There is a long-standing myth that Volkswagens are prone to valve problems," Rex says. "In fact, the valve system is very trouble-free when the heads get enough air and oil to properly cool. Of course, there are other factors that can contribute to overheating, such as an excessively lean fuel mixture and improper timing, but the basic valvetrain will give very good service if not abused.

"We like to use the crankshaft that Volkswagen used on the 1500cc engine in the late '50s. This shaft is a high-quality forging, and has large oil-feed holes to all the bearings, and is not weakened as were later cranks, which have large, cross-drilled journals and elongated oil supply holes that greatly lessened the bearing contact area. The racing crowd found that the older shafts were the most reliable under hard use, and we at HAPI have had no bearing problems whatsoever with this shaft," Rex says. "Perhaps best of all, they are still plentiful and most are standard size, so we can regrind them, taper them for the prob hub, and presto: an inexpensive, available, and reliable crankshaft."

Getting It Together

Having determined what you have, and assuming that your case and crankshaft are good, you are ready to build up your engine, and whether you have previous experience or none at all, you will be well advised to have step-by-step instructions from an experienced VW converter beyond this point. If working with HAPI, you'll need the manual *How to Build a Reliable VW Aero Engine*. It not only details each procedure, but also tells you why things must be done that way. It is well illustrated, and contains all the torque values, clearances, and tolerances.

Preparing for engine assembly, the first thing you must do is take the crankshaft, rods, and pistons to a good machine shop for an operation that you will not be equipped to perform. They will balance these parts both statically and dynamically—meaning that the pistons and rods assemblies will be carefully weighed and all

Fig. 3-12. Taylor's engine manual stresses the importance of proper baffles for effective engine cooling. (courtesy HAPI Engines, Inc.)

machined to match the lightest of each, the weights equalized within a fraction of a gram. Then, an electronic balancing machine will spin the crank and balance the complete assembly. This should be done with the prop hub and flywheel attached. The HAPI manual describes this operation as costing about $60. From there, it is a

Fig. 3-13. Shipment of cases arrives at HAPI from VW. HAPI manual lists all cases that are acceptable for aircraft conversion. (courtesy HAPI Engines, Inc.)

Fig. 3-14. HAPI "Dash-2" series weigh 158 pounds. Note Posa carb mounted beneath the engine, top-mounted oil cooler. (courtesy HAPI Engines, Inc.)

matter of proper engine assembly—and proper installation in the aircraft, not the least consideration being a well-designed cowl and engine baffle system for good cooling (Figs. 3-12 through 3-14).

HAPI is located on the Eloy, Arizona Municipal Airport (about eight miles south of Casa Grande); Rex's wife Phyllis runs the offices and son Pat is in charge of the engine shops (the Taylors also manufacture kits for the Dragonfly and Cygnet homebuilt airplanes). The mailing address is: RR 1, Box 1000, Eloy, AZ 85231. Phone is: (602) 466-9244.

Chapter 4

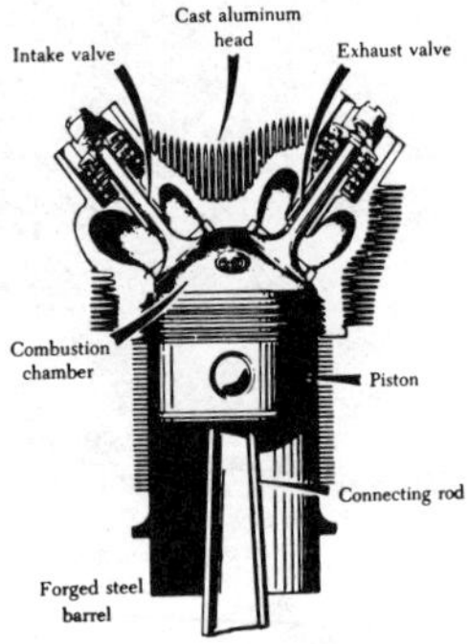

The Duncan Wankel Engines

The Mazda rotary engine was developed from an original design by Felix Wankel of Germany. After years with Mercedes and BMW, Wankel introduced his rotary in 1957, and in 1961 the Japanese firm of Toyo Kogyo (later, Mazda) was licensed to produce the Wankel for automotive use, while in the United States both General Motors and Curtiss-Wright were licensed by Wankel. In Europe, the German Audi firm worked on a lightplane version.

It would seem that there were a lot of heavy guns attempting to develop the Wankel rotary engine, but it appears that C-W's effort was in proportion to the amount of NASA money available to their experiments. The C-W Wankel, installed in a Cessna Cardinal during the early '70s, was a primitive setup employing a dozen V-belts. A later version went into the nose of a Cessna Skymaster (the twin-engine push-pull craft). Neither airplane ever flew with the rotary engine, but C-W was talking big about its RC2-75 (rotary combustion, two rotors of 75 cu/in each) in 1978 at a NASA symposium, comparing the C-W rotary to the Lycoming IO-540. However, when the NASA money ran out, so did Curtiss-Wright's enthusiasm, and the John Deere (farm equipment) company picked up the C-W license.

Mazda

Meanwhile, the Japanese were betting plenty of yen on the Wankel. They offered it in the R-100 and RX-2 to American buyers as early as 1969. The RX-3 and RX-4 automobiles followed, and the Mazda rotary-powered pickup truck entered the U.S. market in 1974, but then, following the Arab oil embargo in 1973 and the fuel shortage scare, the rotary-engined Mazdas were withdrawn from the U.S. market because of their reputation as gas guzzlers and oil burners. Fuel consumption was high, and Mazda had not found a material that would seal the rotor tips and prevent blow-by from the combustion chambers.

The Wankel has no pistons, connecting rods, crankshaft throws, camshaft, or conventional valve and tappet system (Fig. 4-1). Internally, it has a large, triangular-shaped lobe which rotates around an eccentric shaft inside a housing shaped as an epitrochoid (outline of a figure 8). Actually, it is a four-cycle engine—intake, compression, power, and exhaust—but a power "stroke" occurs three times for each revolution of the driveshaft (it can't properly be called a "crankshaft" because it has no crank throws). That makes the single-rotor Wankel equal to a three-cylinder piston engine. The Mazda RX-7 sports car is fitted with a twin-rotor Wankel, and it has the same number of power impulses for each revolution of the driveshaft as a six-cylinder piston engine (Fig. 4-2). Apparently, any number of rotors may be built in series to obtain more power. In Germany, Felix Wankel built a four-rotor engine for test in a Mercedes motorcar.

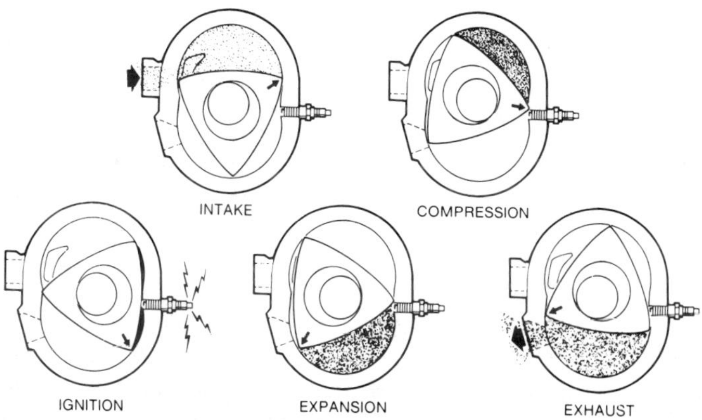

ROTARY ENGINE COMBUSTION CYCLE

Fig. 4-1. A rotary engine generates torque for three-fourths of its operating cycle.

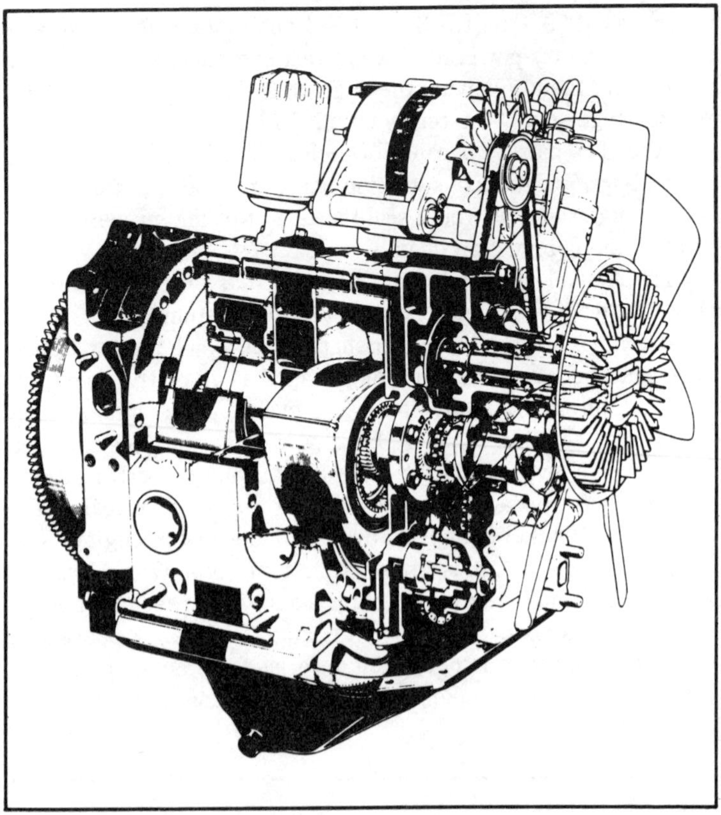

Fig. 4-2. Cutaway of the Wankel dual-rotor engine used in the Mazda RX-7 sports car.

By 1978, Mazda had developed effective apex seals for the engine's rotor, and had dramatically improved fuel consumption. The latter had required some important redesign of the combustion recess in the rotor face, and stratification of the combustion charge. *Stratification* means that a small, rich mixture of fuel and air is ignited by the spark plugs and then is used to set off the major charge, which can be made extremely lean (too lean to be fired by a spark). Because of the unusual configuration of the Wankel's combustion chamber, stratification is more effective than in a reciprocating engine. The stratified charge creates an almost constant flame front through which the rotor continually sweeps the main fuel/air charge. Previously, the specific fuel consumption of the Wankels had been somewhere around .5 pounds per horsepower per hour. Curtiss-Wright was getting .53 sfc with their 285-hp double rotor Wankel.

(Continental and Lycoming recips get down to .41 to .46 sfc.) I don't have the figures for the 1986 RX-7 motorcar, but Dan Duncan claims .25 sfc for his 240-hp twin-rotor Wankel conversions. I could not verify that figure.

Lycoming/Deere Wankel

In the spring of 1985 John Deere announced that it was joining with AVCO Lycoming to develop a series of Wankels for the aviation market. Calling this program "Score III," for Stratified Charge Omnivorous Rotary Engine—"omnivorous" apparently because the Wankel happily feeds upon a variety of fuels, or maybe because it has gobbled up so many good intentions. The joint effort, with $30 million in development funds, is expected to result in both airplane engines and powerplants for John Deere's products. The first Lycoming Wankel is planned at 350 to 400 hp and is being built at the Williamsport, Pennsylvania, plant. Lycoming says that their engine will be controlled by a single lever for throttle, mixture, and propeller pitch adjustment—a concept that Beech worked on back in the '70s and eventually abandoned. Flight testing of the Lycoming Wankel is scheduled for 1987, and production could begin as early as 1990.

Teledyne Continental is also working on a Wankel, said to be an 85-hp twin-rotor known as the R-36. Continental, however, had almost nothing to say about its foray into rotary power at the time of this report.

Duncan Wankel Engines

Duncan Aviation builds up its rotary engines from new cores obtained from Japan, Dan Duncan told me when I visited his facility in Duncan, Oklahoma, early in February 1986. Except for a 40-hp model, the Duncan Wankels are all water-cooled. They are:

Model 40A—27 cu/in, 40 hp. Single-rotor with centrifugal blower, rated at 6,000 rpm. Bosch CD ignition, breakerless, fixed timing; recoil or 12 volt starter, alternator. Recommended time between overhaul (TBO), 1200 hours; operates on regular gasoline, two gph. Weight, 66 pounds dry; 16 inches in length, 14 inches in width, and 14 inches high. With mild steel muffler, $2,000.00.

Model 60W—35 cu/in, 60 hp. Water-cooled, single-rotor, direct drive, 4,000 rpm. Dual spark plugs, Mallory ignition, electric starter (hand props easily); complete with water pump, alternator, and fuel injector. Two prop flange patterns available. Weight, 110 pounds; 18 inches in length, 16 inches in width, and 16 inches high. Fuel consumption, 3.5 gph. TBO, 3,000 hours; $2,495.00 (Fig. 4-3).

Model 120W—35 cu/in, 120 hp. A water-cooled, single-rotor engine with a two-inch HTD Cogg belt propeller reduction drive. Mallory electronic dual ignition; electric starter, alternator, fuel and water pumps; four propeller drive ratios available. Total weight with turbocharger, 165 pounds; fuel consumption, 4.5 to 5 gph; 3,000 hours TBO. Length, 19 inches, width, 16 inches, height, 21 inches. Basic price $3,795.00; turbocharger kit is $1,500.00 (Figs. 4-4, 4-5).

125 to 200 hp—70 cu/in, A water-cooled, dual-rotor version with Cogg belt propeller reduction drive (choice of four ratios); electronic dual ignition, starter, water pump, fuel injection via modified Posa super carb with mixture control or Ellison injection body optional. Alternator; three prop flange patterns. Installed weight, 240 pounds; length, 24 inches, width, 16 inches, height 21 inches; 3,000 hour TBO; fuel consumption, 6 to 9 gph. Price, $4,350.00 (Figs. 4-6 through 4-8).

200 to 240 hp—70 cu/in, Water-cooled, dual rotors, electronic ignition, etc., as above. Installed weight, 280 pounds; 30 inches in length, 18 inches in width, and 21 inches high. Fuel consumption, 9 to 10 gph. Price, $6,500.00.

Fig. 4-3. Duncan Wankel Model SR-60D direct-drive single-rotor engine produces 60 hp at 4,000 rpm. Manually controlled variable-pitch propeller is experimental. (courtesy Duncan Aviation Engines)

Fig. 4-4. Duncan Wankel Model SR-120R produces 120 hp at 9,000 rpm with prop turning 3,000 rpm. Installed weight is 148.8 pounds, plus 16.4 pounds of engine oil and coolant. (courtesy Duncan Aviation Engines)

Aerobatic models are available in 120-hp single-rotor; fuel injected 125 to 240 hp dual rotor, with inverted oil systems.

The 250 to 350-hp turbocharged models are available as aerobatic. Planetary gear reduction units are available on the 60 to 200-hp versions.

Turbocharger kits:	60-120	hp single-rotor,	$1,500.00
	120-350	hp dual-rotor,	1,850.00
Oil coolers and coolant radiators:			$150.00 to 300.00
Constant-speed propeller drive and prop governor drive 125-350 hp models:			$1,000.00

Mazda Cores

Dan Duncan explained that all power increases in the above

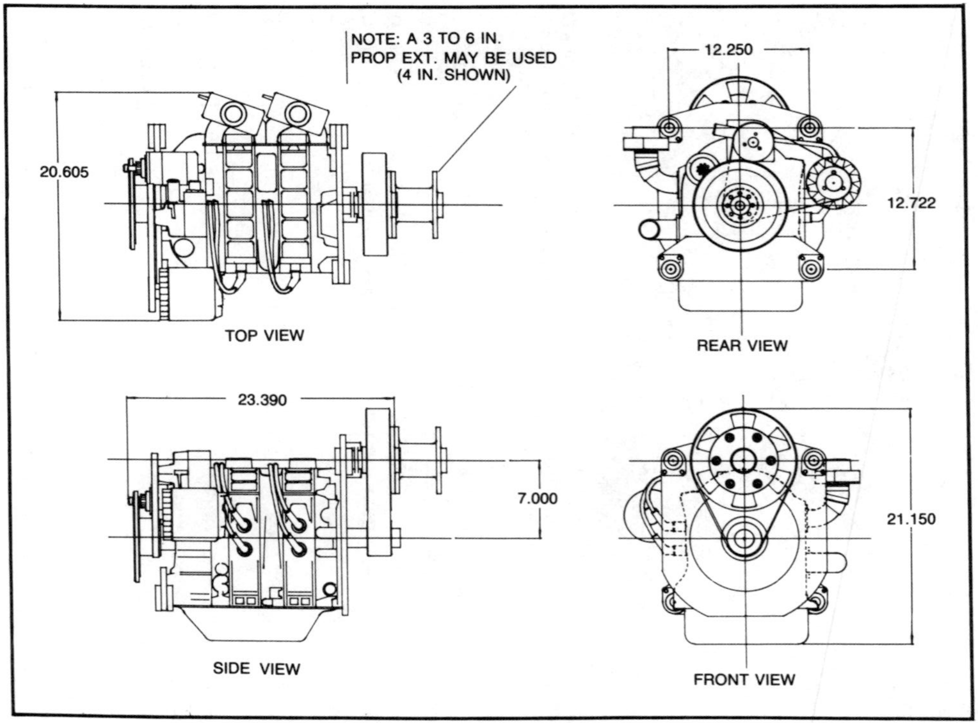

Fig. 4-5. Duncan Wankel Model SR-120R dimensions. (courtesy Duncan Aviation Engines)

Fig. 4-6. The Duncan Wankel DR-200R is a dual-rotor engine which can be rated from 125 to 200 hp; has an installed weight of 240 pounds. On left is the aerobatic version with front support for the propeller reduction drive. (courtesy Duncan Aviation Engines)

Fig. 4-7. The standard reduction drive ratio for the DR-200R aerobatic engine is 2.82:1. This engine produces 200 hp at 8,100 rpm. Maximum propeller rpm is 2,872. (courtesy Duncan Aviation Engines)

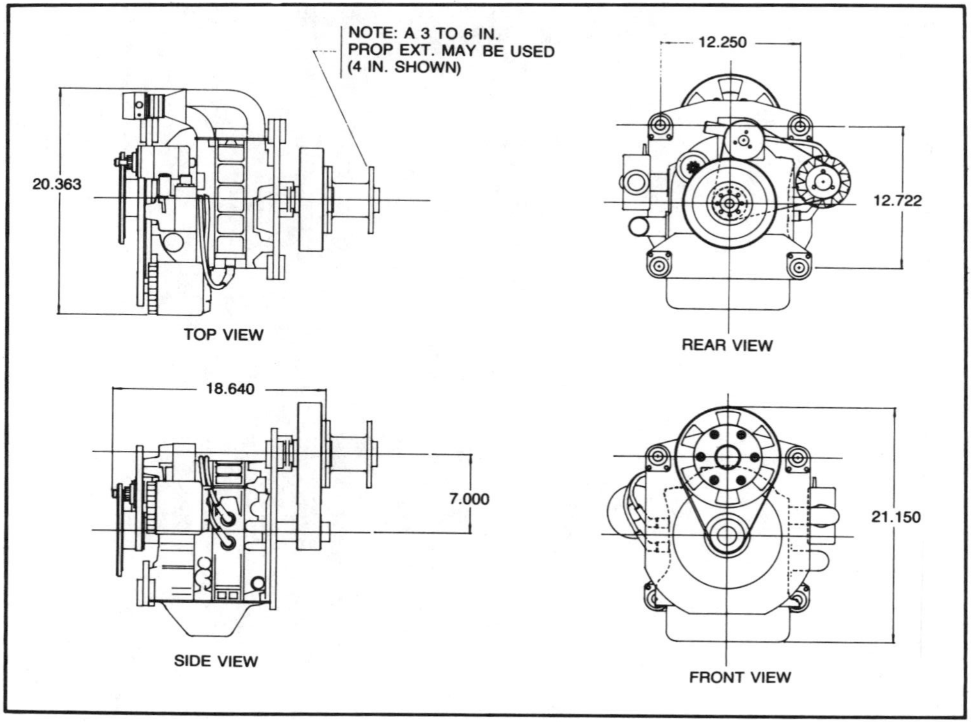

Fig. 4-8. Duncan Wankel Model DR-200R dimensions. (courtesy Duncan Aviation Engines)

engines result from carburetion changes, a free-breathing exhaust, the elimination of emission controls, and increase in rpm. No engine porting or compression ratios have been changed to achieve the power outputs specified.

Evidently, the power increases from the same number of cubic inches comes mostly from more rpm. With main shaft speeds as high as 9,000 rpm, propeller reduction ratios offered by Duncan are 2.50:1, 2.65:1, 2.82:1, and 3:1.

It is easy to get lost in nomenclature when discussing the Wankel engine. I said earlier that this is a four-stroke engine. Duncan adds "ignition" to make it "five phases" or events. A four-stroke piston engine may be said to have the same five events. Anyway, refer to Fig. 4-1, and note the position of the driveshaft through each event. (Duncan calls this the "mainshaft," and Mazda calls it the "eccentric shaft.". Good grief).

1. *Intake:* As the rotor rotates past the intake ports, the swept volume of the chamber is increasing, drawing in the fuel/air mixture.

2. *Compression:* As the apex seal closes off the intake ports, the swept volume of the chamber is being reduced by the motion of the rotor, compressing the mixture. Full compression is reached when the side of the rotor reaches the plug firing position.

3. *Ignition:* When maximum compression is achieved, the plugs fire, igniting the fuel/air mixture.

4. *Expansion:* This is what we'd normally call the "working stroke." After the fuel/air mixture is ignited, the rotor continues to rotate, with the expanding combustion gases exerting force on the rotor face, producing a smoothly applied torque.

5. *Exhaust:* When the swept volume in the chamber is again at a maximum, the exhaust port is uncovered, allowing the spent combustion gases to escape.

We have just followed the process in one of the three working chambers; the other two work in exactly the same way. Therefore, you can see that as the rotor makes one complete rotation, the driveshaft makes three rotations. In other words, there is one power stroke for each rotation of the driveshaft—and the rotor rotates at one-third driveshaft speed.

Rotary Engine Advantages

A single-rotor Wankel has but two moving parts, the rotor and the driveshaft. Since the rotor turns at one-third the speed of the driveshaft, this means that 9,000 rpm on the driveshaft results in a rotor velocity of 3,000 rpm. The rotor constantly turns in the same direction; therefore, the rotor seals do not have to accelerate, stop, and reverse direction as do piston rings in a reciprocating engine. Also, rotary engines cannot be damaged by overspeeding; there is no valve surge or piston ring flutter. Since a Wankel engine has no valves or anything fragile inside, nothing major can break; and if it ain't there, it can't wear.

The driveshafts on these engines are shorter and relatively heavier than those found in recips, and are made of 4130 steel forgings. Duncan says that in the unlikely event that a rotor seal should fail, it would be expelled out the exhaust port, and the engine would continue to run at reduced power. In a twin-rotor engine, all three seals on one rotor could be lost and the engine would continue to produce half power from the remaining rotor. Duncan says that is like having twin-engine reliability with a single engine. The ignition, carburetion, and exhaust systems are independent of one another.

High TBO

"The virtues of having few moving parts, low rotor rotational velocity, low vibration, and liquid cooling combine to make our rotary engines extremely reliable and long-lived," Duncan says. "There is no such thing as a 'top end' overhaul, because there is no 'top end.' Those who are unfamiliar with current Wankel engine technology may be concerned about apex seal life. Mazda resolved that problem years ago by switching from carbon to iron-alloy seals," he continued. "The TBO numbers quoted for the Duncan Aviation Rotary Engines are taken from a careful analysis of Mazda-supplied apex seal wear-rate data. Engine life was calculated at a maximum continuous operating speed of 9,000 rpm and then reduced 25 percent to be on the conservative side. This results in a TBO of 3,000 hours for our normally aspirated and turbocharged liquid-cooled engines."

Test Experience

"All research and development testing, as well as the final checkout of each production engine, is performed on a Flowmaster Digital Dynamometer," Duncan told me. "This dynamometer has a test range of 15 to 1,000 horsepower and measures, records, and/or calculates 17 separate channels of engine data, including horsepower, torque, rpm, fuel flow, air mass flow, mixture ratio, exhaust gas temperature, and various other critical engine parameters. These data are automatically corrected back to standard day conditions from the actual test conditions of temperature, atmospheric pressure, and humidity. The output is printed in tabular form for a permanent record of the test run. This dynamometer has proven to be a most valuable tool for testing the performance of new models and precisely tuning engines . . ."

Maintenance

Dan Duncan told me that standard Mazda parts are used when possible in order to cut replacement costs. They are quality products, in any case, he said. "The Uniroyal HTD Cogg belts used on the propeller reduction drives can be replaced for about 35 dollars," he went on. "Recommended replacement time varies from 500 to 1,000 hours, depending on the horsepower level of the engine. The alternator is a standard Mazda unit, available almost everywhere. A complete major overhaul can be done for less than $500.00. Maintenance largely consists of an oil change and new spark plugs (or clean the plugs) every 75 hours; repack the Timken roller bearings in the reduction drive every 200 hours and, when necessary, changing the belt on the reduction drive takes about 15 minutes. Combine this with the engine's ability to run on fuel as low as 50 octane, and you have a very inexpensive aircraft engine to own and operate."

I consulted the *New Mazda Guide*, TAB book 2082, by Michael F. Hollander, and here is what he reports about rebuilt rotaries:

"Although the current RX-7 engine has increased life over previous rotaries sold by Mazda, periodically the engine will need major repairs. Given the simplicity of the Wankel rotary, there might not seem to be any way to rebuild the engine. But Nick Van Nugteren, owner of Rotary Reliability & Racing, has managed to develop a technique to give the rotary a longer life.

"According to an article in *Automotive Rebuilder* magazine, Mr. Van Nugteren says, 'Rebuilding the Mazda rotary is beautifully simple. Properly rebuilt rotary engines should be capable of operation for 100,000 or more miles.'

"Van Nugteren furnished the expertise for the article and says that rebuilding the engine is merely a matter of measuring parts for wear and reassembling the engine after a complete teardown. There is no 'bore and sleeve' work to be done on the Wankel, so if the housing is worn beyond tolerance, it must be replaced. However, since the other parts are made of softer material than the housing, this is not always a problem . . ."

Fig. 4-9. Dual-rotor Duncan Wankel fitted with VATN turbocharger. The unique design of this turbo unit permits aerobatics. (courtesy Duncan Aviation Engines)

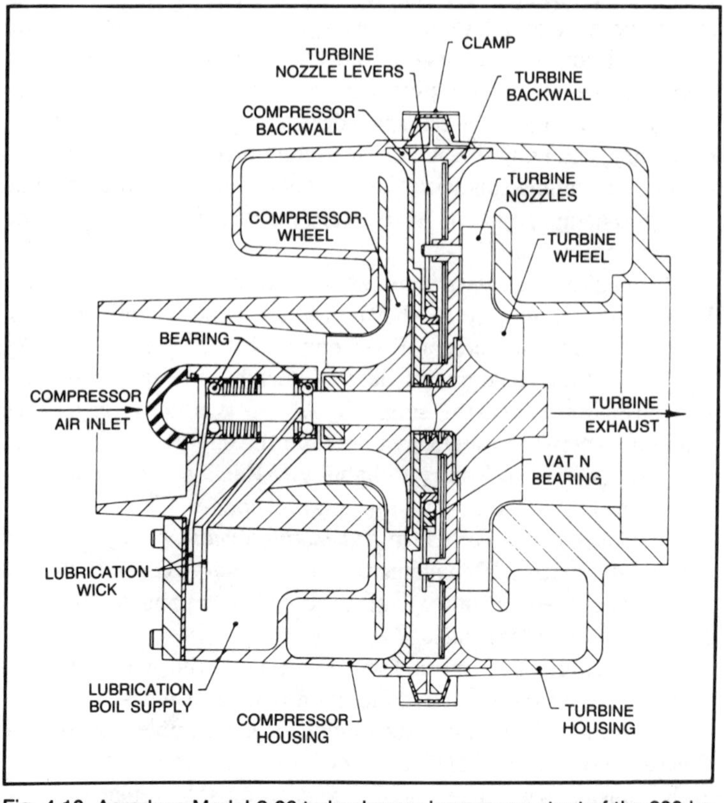

Fig. 4-10. Aerodyne Model 2.66 turbocharger increases output of the 200-hp dual-rotor engine to as much as 350 hp. Single-rotor engines may be boosted to a maximum of 175 hp. (courtesy Duncan Aviation Engines)

The Aerodyne Variable Turbine Area Nozzle Turbocharger

Duncan Aviation is the distributor of Aerodyne turbochargers for the sport aviation/homebuilt airplane market. This unit is unique in three ways: 1) It incorporates a variable area turbine nozzle (VATN) which allows it to adjust turbine nozzle area to maintain peak efficiency and rapid throttle response over a wide range of engine speeds and altitudes. This feature also eliminates the need for a wastegate, with boost pressure being automatically regulated by a pressure-controlled servo; 2) The unit uses low-friction ball bearings located on the compressor side of the turbo, eliminating the danger of hot shutdown damage and extending bearing life; 3) Because of the cool location of the bearings, the turbo has its own self-contained oil supply, requiring no engine oil, and may be mounted in any attitude—and, according to Duncan, flown in any attitude, making it a true aerobatic turbocharger (Figs. 4-9, 4-10).

Wankel-Powered Piper

In the Duncan shops I saw a Piper Tri-Pacer being converted to a sort of Piper Pacer-that is, being transformed into a taildragger, with a long fiberglass nose to house a 250-hp Duncan Wankel. Dan said that he'd have it flying in March. I hope so, because as of early February 1986, I have not seen an airplane fly powered with a rotary engine. It's a real hummer in the RX-7.

Duncan Aviation Engines
500 South Highway 81,
Duncan, OK 73533
Phone: (405) 255-4500

Chapter 5

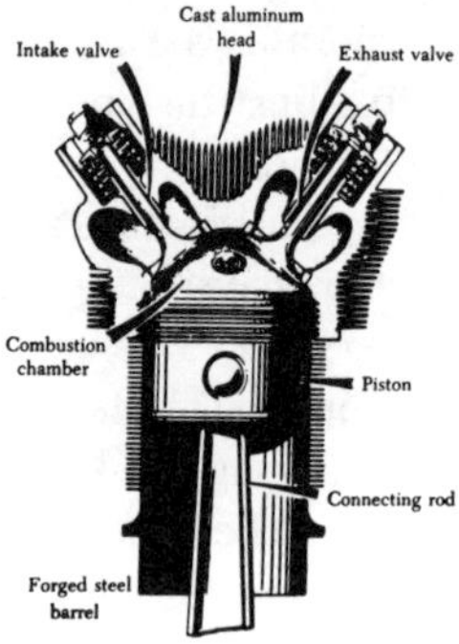

The Revmaster Engines

Revmaster Volkswagen conversions have powered homebuilt aircraft since 1969, and Revmaster's Joe Horvath has, in the meantime, invested a great deal of money and effort in the development of a 2100cc turbocharged engine fitted with a constant-speed propeller. He also offers two normally aspirated engines of 2100cc, rated at 65 and 75 horsepower. The turbo is rated at 80 hp (Fig. 5-1).

Revmaster is located in Hesperia, California, which is almost immediately below as you emerge from Cajon Pass in the San Gabriel Mountains flying northeast from the Los Angeles basin. Interstate 15, leading to Las Vegas, curves around George AFB a dozen or so miles ahead, and then fades into the desert. Edwards AFB is about 40 miles to the northwest. If I lived in California, I'd choose a place like this. I like the desert, and I like people who like the desert.

Aware that there are some basic differences in the way the major engine converters approach this challenge, I asked Joe Horvath about Revmaster's main concerns in converting the VW to aircraft use. His comments are included in Revmaster promotional material, but in person they gain added emphasis:

Fourth Main Bearing

"Starting with the propeller, the gyroscopic forces are tremendous; consequently, a provision must be made to absorb such loads. The small pulley bearing (only 1.9 sq/in of surface area) is not a

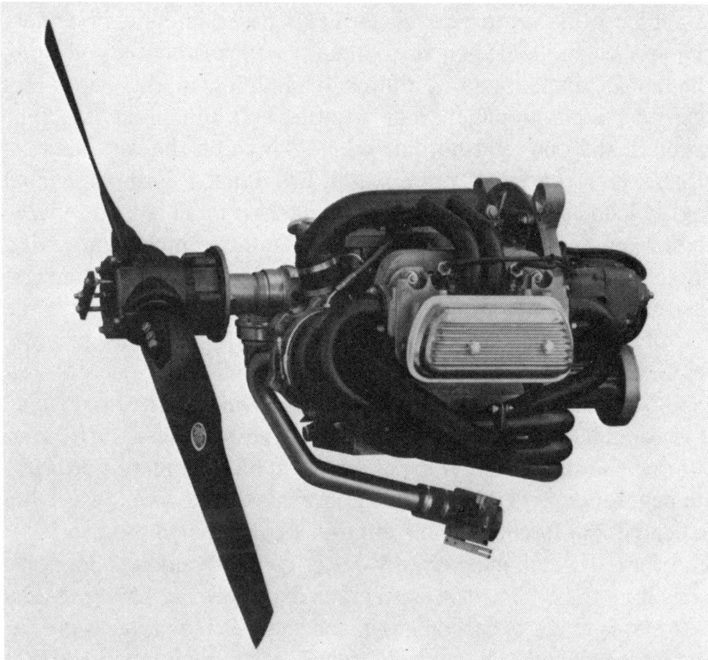

Fig. 5-1. The Revmaster 2100cc VW conversion with turbocharger and constant-speed propeller. Total weight, with accessories, is 209 pounds, and it produces 78 hp at 3,200 rpm and 38″ Hg. (courtesy Revmaster Aviation)

fourth main bearing. Although some claim that the VW has four, it does not. According to the VW company, the engine has only three main bearings. The small front bearing is only a pulley bearing to absorb radial loads of just a few pounds imposed by the fan belt, not hundreds of pounds imposed by the propeller. Without adequate front bearing surface, the loads are naturally transferred back into the center main, which already is marginal.

"By mathematically computing the propeller's gyroscopic force, Revmaster finds it necessary to actually make provision for a fourth main bearing with 13.9 sq/in of surface area. It becomes apparent that without such an arrangement, crankshaft wear will be much greater, including more wear on the main bearings."

Crankshaft

"The second force imposed by the propeller is positive and negative thrust. In the automobile, the only thrust imposed on the crankshaft is the spring pressure of the clutch when depressed. It is not a continuous force such as that imposed in the rear, or near the applied force, designed to take thrust loads before getting to the crankshaft journals. With this in mind, it is quite obvious that when driving a propeller from the other end, the thrust should be taken at the end, and not left where it was on the car. That, of course, is the cheap way of doing it. From this, it is apparent that higher loads are imposed on the crankshaft in an aircraft engine than when used in the car. Since cost is a factor in manufacturing anything, it goes without saying that the car crankshaft is adequate for the car and not with the airplane in mind. Why increase the cost of construction of an engine by making the crankshaft stronger than is required for the application intended?

"Let's take a look at the crank used in most VW conversions. It is a standard, off-the-shelf VW crank." Horvath went on to point out that the standard VW crank has no counterweights, and without counterweights the crankshaft journals tend to fly outward due to centrifugal force, and this imposes high loads on the bearings. The forged Revmaster crankshaft is counterweighted. Horvath sums it up thus: "The standard crankshaft does not have counterweights; is made of carbon steel, and the bearing surfaces are induction hardened. The tensile strength is approximately 100,000 to 110,00 psi. The Revmaster crankshaft is counterweighted, forged of E4340 alloy steel, with nitrited bearing surfaces. The tensile strength is 210,000 to 220,000 psi, and the fatigue life is comparable to type-certificated aircraft engines.

"Also, if the standard automotive crankshaft is used, there is insufficient space to allow a long three-degree locking taper. Instead, a locking keyway is used, which has its limitations. The standard auto crankshaft also does not allow for the large No. 4 main bearing.

"For these reasons, the counterweighted crankshaft as used by Revmaster can have a longer stroke and still not have as great a bearing load as the standard crank with its shorter stroke. In fact, bearing loads mathematically computed for the two cranks establish that the longer stroke of the Revmaster actually has less bearing load than that of other VW conversions.

"There are only a few ways to increase horsepower. One is to increase rpm. This is unacceptable in the case of the airplane, unless gear-driven. The next is to increase displacement. The diameter of the VW cylinder is already at its maximum bore, leaving only two practical alternatives. One is to increase stroke, the other to increase pressure within the cylinder. Because ram tuning is not very effective at the lower rpms, supercharging is the only way to increase BMEP, since the compression ratio is already at its maximum, short of detonation. Obviously, to obtain 75 hp from 1835cc with the same rpm requires quite a lot more pressure to develop the same horsepower as in the 2100cc engine, since the area of the cylinders of both engines are otherwise identical, the difference is a longer stroke.

"Engine reliability and horsepower will determine the success of your homebuilt project. Lack of it can be seen from a few unsuccessful or marginal homebuilts offered on the market today. Figuring horsepower can be necessary and desirable when selecting the engine you are about to convert. One thing to keep in mind is that horsepower in the aircraft conversion will normally be less than this engine produced in the automobile because the aircraft engine will have to turn at a much slower rpm, unless geared or belt-driven. Rpm is dictated by propeller tip speed, which cannot exceed the speed of sound without losing too much prop efficiency.

"The maximum horsepower obtained as an auto engine may be at 4,500 rpm. If this engine is to be used to swing a 60-inch propeller, its maximum rpm will have to be reduced or geared to keep the prop tip within acceptable limits. Therefore, it is prudent to see if your intended engine will be capable of developing the expected horsepower before proceeding with your conversion. This equally applies when purchasing a ready-to-run conversion . . . one way of looking at an engine is to see it as an air pump. It is obvious that the larger displacement will pump more air than the smaller displacement at the same rpm."

Cooling

"In many cases, inadequate engine cooling can be attributed to the airframe designer's omission of engine baffling in the plans, as well as poor inlet and outlet cowling design. Apparently, little thought has been given by many designers in this area, since it has been left up to many of the builders to figure out for themselves. Lack of proper cooling can affect performance to a great extent. It would seem that designers would be more concerned and include better-designed cowlings in their plans. A good design not only adds to performance of the homebuilt, it increases the engine life.

"A good example which illustrates the lack of good cooling design was in the Q-2 cowling and baffling. The outlet was smaller than the inlet, and the inlet was misplaced too far out from the spinner. One must realize that any air entering the cowling that does not come into contact with the engine-cooling fins does not provide cooling but contributes to drag. A second factor to keep in mind is that the outlet must be approximately 22 percent larger than the inlet, since the air, heated during the cooling process, increases in volume. The inlets should be as close as possible to the spinner to obtain a smooth airflow into the inlets.

"An additional factor contributing to poor cooling in the Q-2 was the placement of the exhaust manifolds of the numbers three and four cylinders in front of the engine, allowing heated air to be directed into the cooler. A better design would have had the exhaust running behind the engine. In the case of the Q-2 it was necessary to obtain better cooling by wrapping these exhaust manifolds with some type of insulation material to prevent heat radiating off the manifold into the oil cooler."

Dual Ignition

There are several reasons for considering dual ignition in a VW conversion. The most important is that we are trying to obtain as much horsepower as practical per pound of engine weight. Dual ignition is a way to obtain a horsepower increase while adding little weight. The increase is due to better flame propagation, resulting in more complete combustion, efficiently utilizing more of the fuel.

If you doubt that a dual ignition produces more horsepower, remember the difference in rpm during a magneto check where you have a drop on one system. Take a look at an engine with short stacks and you will see the difference the flame makes when running on one mag compared to running on both. Due to the relative slowness of the flame front, the engine has a longer exhaust flame when running on one mag. This is easily seen at night. It shows that less of the gases are being burned at the time of exhaust valve opening. When this happens, full advantage of the complete charge in the combustion process is not taking place within the cylinder.

Another advantage of dual ignition is that it helps to prevent detonation due to a better flame front. With a single spark plug the flame progresses (especially, at higher power) across the cylinder, the burning gases expanding. This compresses the gases on the far side, which will become hot enough to explode spontaneously. This type of explosion in the cylinder is detonation, a phenomenon injurious to the piston and cylinder head. The remedy is to reduce power and enrich the fuel mixture, if possible.

A third benefit of dual ignition is that it permits a leaner mixture at cruise power without misfiring.

Finally, there is the added safety factor of a redundant system. However, if one mag goes out, power must be reduced and the mixture enriched, if possible, due to the chance of detonation.

Revmaster Turbocharging

Turbocharging offers performance gains for the normally aspirated engine that are not otherwise economically available. Horvath carefully differentiates between "supercharging" and "turbocharging."

"With turbocharging, the energy in the exhaust gases is used to drive the turbine/compressor and provide boost to the intake system of the engine. This energy may be considered as free when compared to a belt or gear-driven supercharger because, if it is not used in this way, it is wasted.

"Many people do not agree," Horvath concedes. "Their position is that exhaust gas energy is not free, because when it is used to power a turbine, that exerts back-pressure on the exhaust system which tends to inhibit good scavenging of the burned gases in the combustion chambers, in turn limiting the volume of the fresh charge entering the cylinder. Well, so does a muffler. However, when an engine is running at maximum throttle setting, with a well-matched, high-efficiency turbocharger, intake manifold pressure will be considerable higher than the pressure in the exhaust manifold. Further, exhaust gas temperatures (EGT) may drop as much as 300 degrees F across the turbine. This drop in temperature represents the fuel energy that is being returned to the engine via the turbocharger. It is a fact that more power, for a given engine displacement and rpm, can be obtained from exhaust gas energy recovery than through any other presently available method

"A turbocharger is a type of supercharger, which is a device that is used to increase the density of the air/fuel mixture admitted into the cylinder. The common synonym for supercharger is 'blower.' It is an air pump that can blow an increased charge of combustibles into the engine's cylinders. A supercharger may be driven by the engine through either gears or belts. This takes power . . . A turbocharger is driven by the engine's exhaust gases, which are expanded across the vanes of a turbine which, through a common shaft, drives a compressor. This blower sucks or blows the air/fuel mixture from or through a carburetor and pushes it into the cylinders. The supercharger is very susceptible to backfiring damage, but the turbocharger is not. The turbocharger also has a decided weight advantage over the gear-driven blower."

Normalized Engines

"A normally aspirated engine running at full throttle at sea level suffers volumetric loss due to intake friction and residual gases left in the space between the piston head and the roof of the combustion chamber. Because of this, most engines will flow only 80 to 85 percent of their theoretical capacity. Thus, running wide open throttle at 29.92 inches Hg (standard atmosphere) is effectively only 24 to 27 inches Hg in the hole where the work has to be done. Next, a compression ratio is created to match the selected fuel at rpm, ignition position, and crankshaft load. Other than tuning the exhaust, what you have at this point is it. This also serves to illustrate why a normally aspirated engine that produces 73 horsepower at sea level (normal day), full throttle, can only deliver something like 54 horsepower full at 10,000 feet pressure altitude.

"Airplane engines that use their exhaust gases to drive a turbocharger to regain loss of power, to altitude only, are known as 'normalized' engines. Traditionally, a normalized engine will have some means of preventing overboosting, or pressurizing the intake manifold beyond design or sea level pressure. Usually, this is an aneroid-controlled bypass valve or 'wastegate,' which dumps all of the exhaust gases at sea level (normal day) and thus none pass over the turbine.

"As the airplane starts its climb and leaves sea level, the engine begins to lose power. To compensate for this, the wastegate becomes partially closed. This shunts a portion of the exhaust gases across the turbine, and the compressor restores the inlet pressure to sea level equivalent. As the airplane continues its climb, the wastegate continues to close; the turbine increases its speed, and the compressor continues to maintain sea level intake pressure. This process continues on up to an altitude where the wastegate is completely closed and all of the exhaust gases are passing over the turbine. If the airplane continues to climb, the engine will start to lose power because the turbocharger can no longer provide sea level intake manifold pressure. This is known as the 'critical altitude' of this engine/turbocharger combination. (We assume that the aircraft design weight/power loading will permit attaining engine critical altitude.)

"The Revmaster premise is that the engine owes us that 15

to 20 percent efficiency loss and that normalizing should begin by giving back to us what we own. Therefore, sea level horsepower at design rpm should be acquired by adjusting the propeller blade-angle to a manifold pressure of 31 to 35 inches Hg at 3,200 rpm. This also means that the observed critical or full-throttle altitude will be lower. However, in the meantime a lot of tall rocks could have been climbed over sooner. Prior experience and data acquired from producing VW-engined race cars and using the Rayjay turbocharger has been very valuable to me. The temperature rise 'ballpark' of the intake system at various mass flow and boost pressures was already known. These data advocated an intake system with a large volume/area trunk and individual cylinder ducting. Placement was on top of, and isolated from, the engine crankcase. This is to acquire maximum cooling effect from the incoming engine cooling air.

"Exhaust manifolding was developed using generous radii for all bends, and collecting at the left aft midsection of the engine. Here, the turbocharger is mounted like the head on an octopus. The system is semi-rigid and is totally separated from the engine by the attachment to the exhaust and intake ports of the cylinder heads and no other support is required.

"Now, as you supercharge the engine, obviously the heat release in the combustion chamber goes up because there is more fuel being burned in a given unit of time. So supercharging always increases the cylinder head temperature because the gas inside is more dense, and therefore the heat transfer into the cylinder head is greater. Average density is greater; heat flux is proportional to density, among other things, and heat flux out of the cylinder head is more or less constant. So the temperature has to rise in order to get the heat to flow out of the cylinder head using the same cooling potential. It just has to be done at a higher differential temperature. This shows up on the meter as a higher structural temperature of the cylinder head. At high compression ratios the pressure and temperature increase that results from initial combustion may cause detonation as previously mentioned. Revmaster does three things to alleviate this condition: 1) We use a piston with an offset wrist pin, which favors rotation of the crankshaft; 2) The compression ratio is tailored to the end usage, and 3) Only 100 octane fuel is used."

Oil Cooling

A lot of 2100cc Revmaster engines were sold without turbocharging, and this normally aspirated engine continues to be available. With the advent of the turbocharger and Maloof/Revmaster variable-pitch propeller, it was necessary to reexamine the heat-rejection capability of the basic R-2100 engine. Heat rejection of this engine is parceled out three ways:

- ☐ Air exchanged across the cylinder heads and, to a lesser degree, the cylinders.
- ☐ Engine exhaust. Approximately one-half of the heat of combustion is not recovered during the engine expansion cycle.
- ☐ Circulation of the engine oil through the crankcase, the exposed cylinder walls, and through the heads to lubricate the valvetrain. All of these areas add heat to the oil.

The oil rejects this heat either directly through the crankcase walls or through an auxiliary cooling device—an oil cooler. It is important to remember that the heat-conducting path available to the oil is usually very massive, while the heat flow is generally outward through the crankcase walls. Heat transfer out of the crankcase can be very limited. This is because the only way it is cooled is by handing off its heat energy to the air and, to a small degree, by radiation to the cowling walls. Unfortunately, most cowling walls are reflective to a high degree, so much of this radiated heat gets focused back onto the engine.

As the heat load from the engine increases, the crankcase eventually gets to a point where its temperature stabilizes. This may be at a temperature that is too high for the oil. The turbocharger operating at design speed (85,000 to 100,000 rpm) heats its lubricating oil to 290 to 300 degrees F. This flow may exceed one-half gallon per minute. With an engine oil body of four quarts, this becomes significant. There are two ways to improve the situation. One is to direct more outside air onto the crankcase. The other is to divert the oil body into a heat exchanger or oil cooler. Actually, both of these things must be done in order to have a weight-efficient system.

It should be borne in mind that the amount of power that an engine is normally capable of developing varies with altitude; also, that the heat-rejection capacity of the air varies in the same way. So normally-aspirated engines that tend to cool adequately at sea level also tend to cool adequately at altitude. What can change that happy state of affairs is the addition of supercharging. Then the heat rejection remains more or less constant with altitude while the air density and, therefore, the ability of the cooling air to carry the heat away, declines with altitude. Supercharged engines tend to run somewhat hotter at altitude for that reason. Counteracting that is the fact that a heat exchanger's efficiency varies approximately as the third power of the temperature differential, i.e.: the difference in temperature between the fluid coming in and the temperature of the cooling air coming in. If it were pure radiation, the advantage would be even higher, because the radiation efficiency varies as the fourth power of the temperature differential. The exact component cannot be determined without measurement. Nevertheless, there is a dramatic improvement in the efficiency of the oil cooler with small increases in temperature differential.

Aside from the thermal efficiency of an oil cooler, there is a resistance-to-flow factor. This precludes the use of the automobile-type coolers. Many of these types generate a pressure drop that seriously inhibits cooling and lubrication to the engine. Their size, shape, and weight, plus mounting and drag problems, have led to the development of a two-header, finned tube, large-area unit that is mounted beneath and parallel to the bottom of the crankcase. This mounting has minimum cooling air drag, as well as minimum hose lengths or tube attachments. Weight of this exchanger is 1.5 pounds, and the pressure drop (hot oil) is three to five pounds, minimal.

Cylinder Head Temperatures

What you measure under the spark plug is the *average* temperature of the cylinder head and not the *instantaneous* temperature. The mass of the cylinder head tends to level out the temperatures and make the variations in temperature with each explosion cycle undetectable. They are there all right, but you can't measure them. So increasing the compression ratio of the engine will increase the high and low surface temperature at the inside surface of the combustion chamber. What would be detected, short of detonation, would be a level temperature lower for a higher compression ratio.

As the compression ratio goes up, the temperature of the explosion wave also goes up, and that increases the temperature of the top of the piston. The mean temperature of the cylinder head actually goes down as the compression ratio is increased, providing that you do not get into detonation. Detonation, at any compression ratio, will increase the heat load because the explosion phenomenon represents a decline in engine expansion efficiency. That is what the explosion wave does when it gets past a certain density. In diesel engines this is controlled by admitting the fuel progressively so that the charge temperature tends to remain constant for a time and at a somewhat lower temperature. The most efficient engine is one with the highest achievable compression ratio (short of detonation) because more of the heat of combustion is converted to useful work. This also means that less of the heat of combustion is converted to heating the engine structure and less is left to heat the exhaust gases. The coolest running exhaust is always from the engine with the highest compression ratio.

Compression Ratio and Turbocharging

There is a tradeoff in compression ratio and fuel/air ratio for turbocharged engines because you must have enough exhaust energy to operate the turbine, so the maximum compression ratio does not necessarily give you the best turbine operating environment, Revmaster says. That would come at a lower ratio but with the overall efficiency of the engine reduced. You will be able to acquire more horsepower, but it would be less efficiently produced. There is also a gain in horsepower produced as altitude increases because the back-pressure seen by the exhaust decreases. This amounts to as much as seven percent at 14,000 feet. The optimum compression ratio depends on the drag characteristics of the airplane in which the engine is to be mounted the speed at which it is to be flown, and the altitude. The lower and slower the aircraft, the lower the engine compression ratio using a turbocharger. The normally aspirated R-2100 Revmaster engine has a compression ratio of 9.35:1, while turbocharged the ratio is adjusted as required to between 7.1:1 and 7.8:1. One hundred octane fuel is used in all Revmaster engines, all compression ratios.

Revmaster Propeller System with Turbocharger

The Revmaster turbo propulsion system consists of the R-2100-D engine, Maloof oil-controlled constant-speed propeller (Fig. 5-2), turbocharger, injector carburetor, 20 amp alternator, starter, oil cooler, and oil filter. This combination weighs 209 pounds. Power developed is 78 hp at 3,200 rpm and 38 inches Hg.

Overboost?

Special safeguards have usually been contrived to try to pre-

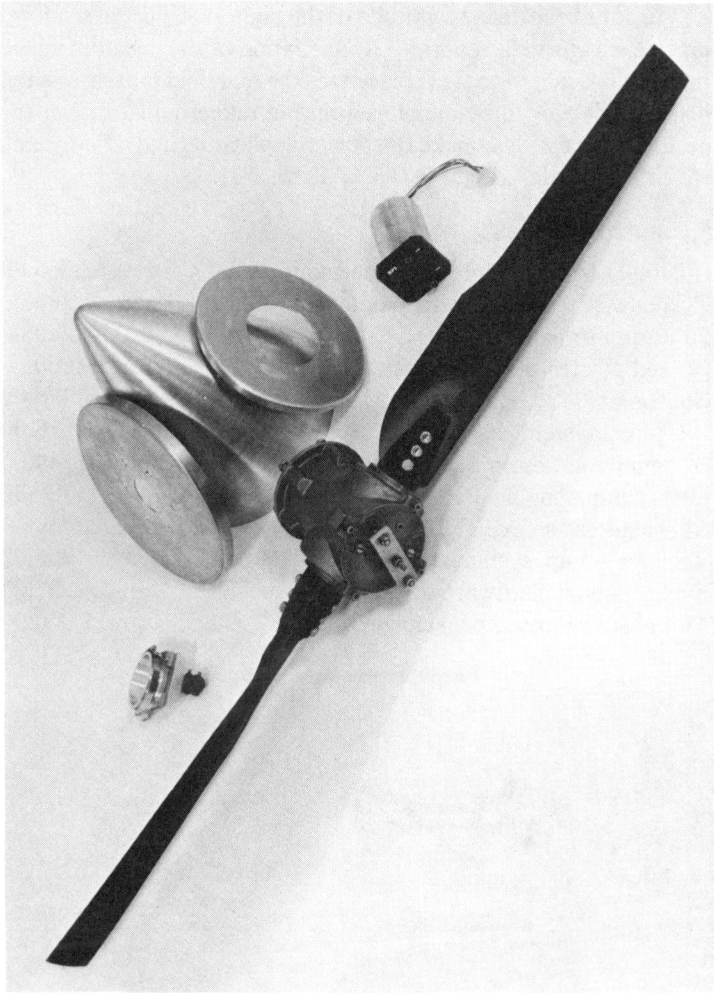

Fig. 5-2. The Revmaster-Maloof constant-speed propeller is oil-controlled. (courtesy Revmaster Aviation)

vent engine abuse from excessive boost pressure. Some of these are exhaust wastegate, water injection, mixture enrichment, ignition retard, and intake pressure or flow-restricting valves. Revmaster takes the position that the most reliable overboost protection is an "informed and alert pilot." Therefore, the Revmaster system uses the throttle as the boost pressure limiting device.

Revmaster says that pilot acceptance has been good. "Particularly rewarding has been the comment on the ability of this system to adjust the density altitude of the highest airport to sea level pressure," Horvath reports. "Appreciation of the constant-speed function is quick to come. This leaves the manifold pressure as the master indicator to be monitored during takeoff. Just dial in the desired rpm and then make the power application and adjustments with the throttle" (Figs. 5-3, 5-4; Table 5-1).

Revmaster Prices

R-2100-D basic engine with dual ignition	$2,965.00
Turbocharger system with plumbing	1.400.00
20 amp alternator	220.00
Geared starter and ring gear assembly	260.00
Oil cooler	126.00
Oil filter system	78.00
Oil sump side drain assembly	44.00
Fuel pump, engine-driven	120.00
Alternate air source/air filter assembly	45.00
Vacuum pump system	325.00
Engine mount hardware kit	27.00
Altitude performance package	1,000.00

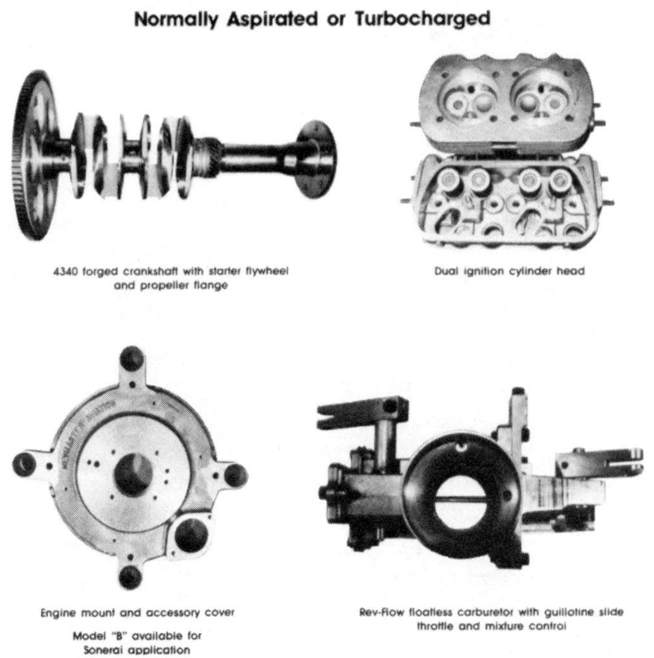

Fig. 5-3. Revmaster engine components. (courtesy Revmaster Aviation)

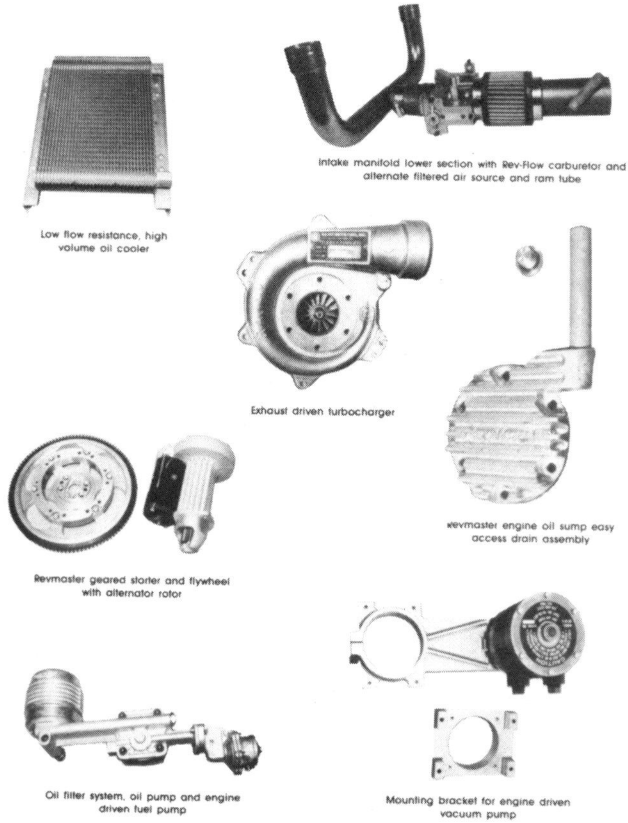

Fig. 5-4. Revmaster engine components. (courtesy Revmaster Aviation)

Rev-Flow carburetor with mixture control (separate order)180.00
Maloof/Revmaster Propeller, constant-speed 1,700.00
Maloof/Revmaster Propeller, two-position 1,420.00
Write or call Revmaster for sales terms and warranty.

System #1—R-2100-D, 65-hp normally aspirated engine including E4340 forged crankshaft, dual ignition, mixture control, intake manifold, 12 volt starter and ring gear assembly, 20 amp alternator, voltage regulator, oil cooler, oil filter system, oil sump side drain assembly, alternate air source and filter assembly, shipping legs and carton. FOB Revmaster $3,751.00

System #2—R-2100-D, 75-hp normally aspirated engine including the E4340 forged crankshaft, etc., as above, plus special altitude performance cylinder head and intake manifold package, shipping legs and carton. FOB Revmaster $4,751.00

Table 5-1. Revmaster Engine Specifications.

The Revmaster aircraft engine models described herein meet the self-imposed Part 33:00 simplified engine criteria for light aircraft and JAR 22 (April 1, 1980) European joint airworthiness requirements for powered sailplanes and motor-gliders.

	2100-D N/A 65	2100-D N/A 75	2100-D Turbo 80
HP Rating	65 @ 3200	75 @ 3200	80 @ 3200
Max continuous hp (standard day)	65 @ 3200	65 @ 3200	70 @ 3200 32″hg
RPM sea level pressure alt. take-off hp 5 min. rpm (full throttle)	65 @ 3200	75 @ 3200	80 @ 3200 36″hg
Fuel minimum grade	100 octane	100 octane	100 octane
Lubricating oil (multi-grade)	20/50	20/50	20/50
Bore & stroke	cc 78 × 92	78 × 92	78 × 92
	in. 3.070 × 3.622	3.070 × 3.622	3.070 × 3.622
Displacement (cu. in.)	2074	2074	2074
Compression ratio	9.34:1	9.34:1	8.5:1
Weight (dry)	165	165	196
Prop mnt. flange SAE	#1	#1	#1
Carburetion	Rev-Flow 32	Rev-Flow 32	Rev-Flow 32
Ignition	Bendix D-3000	Bendix D-3000	Bendix D-3000
Timing BTC	25°	25°	25°
Spark plugs	See notes		
Oil sump capacity	3.5 qts.	3.5 qts.	3.5 qts.
(with C/S prop & oil filter)	4 qts.	4 qts.	4 qts.

System #3—R-2100-D, 75-hp normally aspirated engine including same as System #2, with the addition of Maloof/Revmaster two-position propeller. FOB Revmaster................$6,171.00
System #4—R-2100-D, 80-hp turbocharged engine including E4340 forged crankshaft, dual ignition, Rev-Flow carburetion system with mixture control, intake manifold, 12 volt starter and ring gear assembly, 20 amp alternator, voltage regulator, oil cooler, oil filter system, oil sump side drain assembly, alternate air source and filter assembly, turbocharger system with plumbing, Maloof/Revmaster constant-speed propeller (52″ to 56″), with combination tachometer and propeller speed controller, and shipping carton. FOB Revmaster........................$6,851.00

Revmaster Aviation
P.O. Box 2084,
Hesperia, CA 92345
Phone: (619) 244-3074 TELEX 676459

Revmaster R2100-S

The Revmaster R2100-S is a 2100cc, 7.22:1 compression ratio, 87 octane auto fuel basic engine with single ignition, 62.5 hp rated at 3200 rpm, including the following:

- ☐ E4340 forged, counterbalanced crankshaft
- ☐ E4340 forged, gold anodized prop flange
- ☐ Massive #4 main bearing - 13.5 sq. in. of surface area to take up imposed propeller thrust loads
- ☐ Aluminum finned valve covers
- ☐ 92mm steel cylinder set with forged pistons
- ☐ Rear engine flywheel cover and engine mount housing
- ☐ Flywheel and ring gear assembly (to accommodate optional geared starter)
- ☐ Gold anodized intake manifold assembly
- ☐ 32mm Rev-Flow carb with built-in mixture control
- ☐ Revmaster alternate air source (carb air filter)
- ☐ High capacity oil pump and oil filter assembly
- ☐ Tube and fin "full flow" oil cooler assembly with stainless steel braided lines and aluminum anodized fittings
- ☐ Oil sump side drain assembly
- ☐ Single ignition magneto or capacity discharge ignition system

The R2100-S, ready to ship in carton with shipping legs, FOB Hesperia, California, is priced at $3,055.00.

Chapter 6

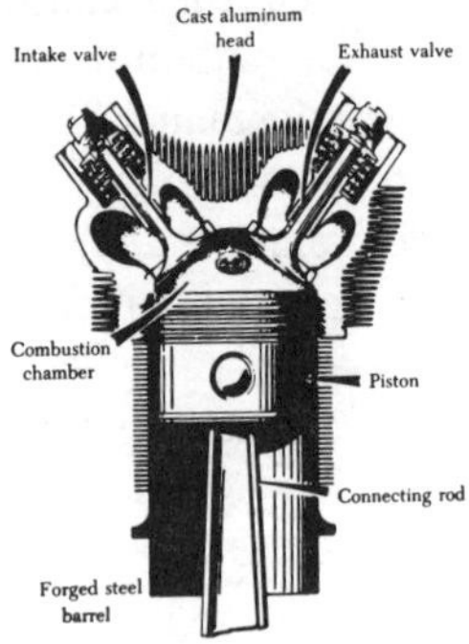

The Limbach Aircraft Engines

Most amateur airplane builders in the United States have heard of the German-built Limbach engines. Peter Limbach and son Peter, Jr. have been generous contributors to the Experimental Aircraft Association for years, although they did not sell engines in the U.S. until late 1984. These engines, ranging in power from 60 to 100 hp, have dominated the European motorglider scene for more than a decade, but the Limbachs were skeptical of what they perceived to be the Americans' propensity to tinker with things, and they didn't want their reputation jeopardized by events beyond their control.

My link with the Limbachs has been J.C. Treager, General Manager of Limbach Aircraft Engines, with offices in the terminal building on Tulsa International Airport. Trained as an engineer, Treager was a soldier by profession (now retired), and admits to a lifelong love affair with airplanes. He maintains an inventory of the most popular Limbach models at Tulsa.

The best way to describe the Limbachs is to say that they evolved from the Volkswagen. "We are a little sensitive about referring to our engines as VWs," Treager says. "Certainly, they were born out of the VW 'Bug' era. We are in fact proud of that heritage . . . what we do not want to happen is to have our engines compared to VW auto engines converted to aircraft use.

"Our engines, particularly the Model SL1700 series, do have a number of VW parts; however, they have been procured new by

Limbach and integrated into a *designed* aircraft engine. Those engines have carried German government type certification as commercial motorglider engines for many years. The point is, they are produced in a *certified* environment which adds a level of confidence for users . . .

"Our main production engines as used in the modern motorgliders range from 75 through 90 horsepower. Except for the case, all essential components are of Limbach design, and custom-manufactured in Germany—for example, forged crankshafts and forged pistons. Even the cases, obtained from Volkswagen, are highly modified by Limbach to accommodate their unique aircraft design."

Crankshaft Bearing and Other Questions

I asked Treager about the front main bearing where, it seemed to me, most of the propeller loads would be concentrated.

"There have been about 3,000 Limbach aircraft engines put in use over the past 20 years," he replied. "After hundreds of thousands of hours of operation, these engines have not experienced any thrust bearing problems. It may help to repeat Peter Limbach, Jr.'s answer to that question. Peter Jr. is the company's chief engineer and designer of these engines: 'We are asked quite often what we have done about the thrust bearing. Our answer is quite frank; nothing. Let me explain.'

"'Most of you probably know that our engines have a derivative linkage to the Volkswagen air-cooled engine design. All of the air-cooled VW engines have a thrust bearing at the rear which takes the load applied during clutch action. The crankcase has been designed to take such loads at exactly that location. We therefore chose not to relocate the thrust bearing simply because the crankcase is not designed to take axial loads anywhere else.

"'The actual static thrust loads the bearing surface with approximately 57 psi, which drops to a mere 14 psi under cruise conditions. These are very low figures such that bearing life should be no problem at all. That we have several thousand engines in operation should mean we are doing something right.'"

Limbach Engine Mounts

Reminded that most homebuilts, particularly the kitplanes, will require specially designed engine mounts in order to fit a Limbach engine, I asked Treager about that. He had already addressed that question, he said, in the Limbach Newsletter. He added:

"For several reasons, we have not attempted to provide engine mounts and other airframe accessories. We prefer that the supply of such items needed for custom aircraft be furnished or fabricated by either the builder or supply firms that cater to the custom built aircraft community. Basically, we make 'generic' aircraft engines that are suitable as powerplants for a wide range of commercial and custom designs. That's the business we know, and it is the business prudence guides us to stick with.

"However, we do want to help those of you who want to use a Limbach engine and need help with the mount. To that end, we are cooperating with an EAA friend who has successfully operated a small supply business for many years, providing special components to homebuilders. You have no doubt heard of Starfire Aviation and its owner, Mac McKenzie. Mac is best known as the guiding force behind the refinement and popularity of the Skybolt biplane design. Mac is tooling up to produce on order Limbach engine mounts for several of the popular kitplanes. Contact Mac directly if you have an interest in one of his Limbach engine mounts." That is: Mr. H.G. McKenzie, Starfire Aviation, 910 South HoHoKam Drive, Unit 107, Tempe, AZ 85281; Phone: (602) 968-2556.

Limbach Overhaul

To my question about engine overhaul, Treager said, "I had the tough assignment of spending two weeks at the Limbach Flugmotoren plant during the last half of September 1985. It is always a pleasure to meet with friends and partake of the beautiful scenery and great food that each trip to Germany brings. An added treat this trip was to have my close friend, Bud Upton, accompany me. Bud is a well-known pilot and master airplane mechanic from Tulsa. He is an ex-president of EAA Tulsa Chapter 10 and always active EAAer. Bud has every rating from Airline Transport Pilot to AI mechanic. He is also the FAA designated examiner for A&P candidates. He went to our factory to receive intensive training on all facets of the Limbach engines. Bud is now available to perform almost any kind of repair on these engines, including accessory rebuild. You can contact Bud at our Tulsa office."

Fuel, Etc.

Can Limbach engines use automobile fuel?

"Yes. However, since all of our certification testing was done in Germany using aviation fuel, legal considerations demand that we recommend that only aviation fuel be used. We know many Limbach engines are run on auto fuel, and that homebuilders will no doubt do so. We have not tested any auto fuels in our engines; therefore, if you use auto gas, you do so at your own risk. Herr Limbach believes that aircraft using auto fuel should be specially equipped with separate fuel pumps to ensure against vapor lock problems."

In what custom-built aircraft designs can Limbach engines be used?

"We believe that the 68 through 90 horsepower engines will be suitable for most light/sport aircraft. We especially have had many Dragonfly, Sonerai, and Polliwagen builders indicate interest in our engines. We are working with builders in each of these groups to develop standard installation packages."

What are your recommendations concerning propellers for Limbach engines?

"Our chief engineer believes that the engine-propeller match is critical to the successful service life of any aircraft engine. All of our engines used in commercially produced motorgliders presently use the very high quality Hoffman fixed and variable-pitch propellers made in Germany. We definitely recommend—*insist*—that you use a quality wooden propeller with our engines. We are considering a plan to have a standard 'EAA' propeller made in Germany to use with the engines for homebuilders. We would make such a prop available at a very attractive price—near cost—as an inducement to use it.

There are quite a few motorgliders in the United States that have Limbach engines. Can the owners get parts and other support from you?

"Absolutely. We very much want to hear from and support all those loyal owners. Also, any engines that are near overhaul time may be traded up to higher power. In any case, contact us about replacing instead of overhauling. You may be able to get a new one for little more—or less—than the U.S. overhaul price."

The Company

The Limbachs are manufactured in Sassenberg (near Bonn) Germany, and Herr Peter Limbach, and son Peter, Jr., do acknowledge a certain debt to Volkswagen because today's Limbach Flugmotoren was originally Limbach Motorenbau, a company which turned out more than 25,000 rebuilt VW engines for Sears and Montgomery Ward back in the '60s. As a result of that business, the Limbachs encountered virtually every type of failure possible in those engines, and that proved a useful experience when the company turned to aircraft powerplants.

The change began in 1968, and was sparked by a request from Sportavia, the company that builds the French-designed RF-4 and RF-5 motorgliders in Germany. Sportavia was hoping that Limbach could get more power from the 1200cc VW engines fitted to their machines. The result was the Limbach SL1700E ("SL" for Sportavia Limbach, 1700cc displacement, and "E" for *Ein*, single ignition).

The SL1700E1, which has Type Certificate 4582 from the German equivalent of our FAA, produces 60 hp at 3,000 rpm, is single ignition and, at this writing (February 1986) is priced at $3,300.00.

The Limbach factory is currently building two types of aircraft engines and a ground power unit, the latter used by German fire departments. The aircraft engines are four-stroke, four-cylinder in the 60 to 100-hp range, and two-cycle engines in the 22 to 50-hp range. The two-cycle engines are sold to several governments for use in military drones.

Motorgliders

The Limbachs, and their U.S. General Manager, J.C. Treager, call our attention to a form of sport and personal flying that is just now beginning to attract a significant number of U.S. pilots. The motorglider, as it has evolved in Europe, is sort of misnamed. It is almost exclusively "motored," seldom glided. We've had pure motorgliders—or at least, self-launching sailplanes—in the United States for many years, and the Nelson two-cycle engine was actually FAA certified for aircraft use as a result of the anticipated market for an auxiliary powerplant that would allow sailplane pilots to dispense with the cost and bother of a towplane. Most such installations were envisioned as retractable—engine and prop folded down into the fuselage or power pod once the sailplane reached its desired altitude or found lifting air. A worthwhile market for such a combination did not develop.

At this writing, there is but one U.S. sailplane manufacturer—or, more accurately, a producer of kits. That is Monnet Experimental Aircraft (Box 2984, Oshkosh, WI 54903; phone: 414-426-1212. Scheweizer Aircraft of Elmira, NY, produced gliders and sailplanes for decades, but appear to have stopped production). The Monnet Monerai (Mon-a-ray), with a lift/drag ratio of 32:1, is not in the same class with the German fiberglass competition machines performance-wise, but it is not in their price class, either. The Monerai kit is priced at $4,150.00, while the German super soarers start at about $60,000. But those are pure sailplanes, and the purists in the movement turn up their noses at the thought of adding a smelley ol' engine to one of their pristine birds.

Of course, the sailplane people turn up their noses at most things. They are an arrogant and obnoxious clan.

The Monnet Monerai falls short of the current European motorglider concept because its engine—which costs $2,450.00 extra—is a 25-hp two-cycle that is quite adequate for self-launch, hopping between thermals, and providing a choice of landing sites, although less than ideal for sustained, powered flight. And while some high-performance sailplanes equipped with anemic engines are being made, the hottest sellers these days are the two-place, soarable motorgliders that are flown primarily in the power mode. In Germany today, most of the hours flown by the motorgliders are as powered airplanes being used for sport and/or private transportation.

In Europe, motorgliders are very popular as club and rental aircraft. A typical two-place motorglider will cruise at 120 mph on 2.5 gph (automobile unleaded), will have a 700 fpm initial climb rate with two big people aboard, and its wings are foldable for home storage.

The first motorglider rally in the U.S. was held at the Hemet-Ryan Airport (near Riverside) on 18 and 19 January, 1986.

Limbach SL1700E and E1

Four-cylinder, horizontally opposed air-cooled; wet sump, pressure lubrication, single ignition (Figs. 6-1, 6-2).

Maximum hp (5 min)......................68 hp/3550 rpm
Continuous hp..........................60 hp/3000 rpm
Bore...............................88mm/3.46 inches
Stroke.............................69mm/2.71 inches
Displacement........................1680cc/102.52 cu/in
Compression ratio.................................8:1
Magneto....................Slick 4230 or Bendix S4RN21
Carburetor............................Zenith 150 CD3
Starter with solenoid.......................12 V - .5 KW

Fig. 6-1. The Limbach SL1700-EA1 rated at 60 hp at 3,550 rpm weighs 154 pounds, is type-certificated by Germany's LBA, equivalent to the U.S. FAA. (courtesy Limbach Aircraft Corp.)

Dry weight . . . . 73 kg/160.6 lbs
Fuel consumption . . . . 10 ltrs/hr, 2.27 gal/hr

Propeller hubs available for variable-pitch Hoffman HO-V 62R propeller, or fixed-pitch Hoffman HO 11.

Fig. 6-2. Frank Kopecky of Van Nuys, California nears completion of his Limbach-powered Dragonfly. Engine is the SL1700-E version of 68 hp. (courtesy Limbach Aircraft Corp.)

Limbach L2000DD

Four-cylinder, horizontally opposed air-cooled; wet sump, pressure lubrication, dual ignition (Figs. 6-3 through 6-7).

Maximum hp (5 min) . . . . 76 hp/3400 rpm
Continuous hp . . . . 68 hp/3000 rpm
Bore . . . . 90mm/3.54 inches
Stroke . . . . 78.4mm/3.09 inches
Displacement 1994cc/121.68cu/in
Compression ratio 8.3:1
Magnetos . . . . two Slick 4201
Carburetor . . . . Zenith 150 CD-3
Starter . . . . Bosch 12 V - 0.5 KW
Dry weight . . . . 75 kg/164 lbs
Fuel consumption . . . . 12 ltrs/hr or 3.17 gal/hr

Propeller hubs available for Hoffman variable or fixed-pitch, or SAE Propeller Hub #1.

At the time of this writing—February 1986—the L2000DD engine had not yet received its type certificate from the German Government. Following is a complete list of Limbach engines, with current prices:

Certificated

SL1700 E and E1 . . . . . 68 hp/3550 rpm . . . . . 73kg/160.6 lbs
SL1700 EA and EA1 . . 60 hp/3550 rpm . . . . . 70 kg/154.0 lbs

Fig. 6-3. Limbach L2000-DD1 is rated at 76 hp at 3,400 rpm; 68 hp at 3,000 rpm cruise. Dry weight is 164 pounds. (courtesy Limbach Aircraft Corp.)

SL1700 EB and EB1....72 hp/3600 rpm.... 72.5 kg/159.5 lbs
SL1700 EC and EC1....68 hp/3600 rpm......73 kg/160.6 lbs
SL1700 EAA and EAA1.56 hp/3550 rpm. 60-68.5 kg/132.0 lbs
L2000 EO1............80 hp/3400 rpm.... 72.5 kg/159.5 lbs
L2000 EA and EA1.....80 hp/3400 rpm......69 kg/151.8 lbs
L2000 EB1, A, B, C....80 hp/3400 rpm......73 kg/160.6 lbs
L2000 EAB1...........80 hp/3400 rpm......73 kg/160.6 lbs
L2000 EC1............80 hp/3400 rpm.... 72.5 kg/159.5 lbs
L275 E (drone).........22 hp/7200 rpm...... 6.5 kg/14.3 lbs

Non-Certificated

L1700 D, DA, DD...... 65 hp/3600 rpm..... 76 kg/167.2 lbs
L2000 D, DA, DD...... 76hp/3400 rpm...... 75 kg/165.0 lbs
L2100 E, 1X........... 74 hp/3200 rpm... 72.5 kg/159.5 lbs
L2300 E, B1...........85 hp/3200 rpm..... 81 kg/178.2 lbs
L2300 D, DA, DD......86 hp/3200 rpm..... 87 kg/191.4 lbs
L2500 EB1............90 hp/3200 rpm..... 81 kg/178.2 lbs
L2500 D, DA, DD......90hp/3200 rpm...... 87 kg/191.4 lbs
L2700 E, D, DD........90 hp/3000 rpm...93-96 kg/204.6 lbs
L3000 E, D, DD........100 hp/3200 rpm..93-96 kg/204.6 lbs
L550 E (drone, Fig. 6-8).50 hp/7200 rpm...... 17 kg/37.4 lbs
L200 A (ground power)..48 hp/2500 rpm.... 100 kg/220.5 lbs
L90 E (experimental)....6 hp/8500 rpm....... 4.5 kg/9.9 lbs

Prices

Engine Type	*T/O Horsepower*	*Price in $ U.S. (ex factory)**
Sportavia/Limbach SL1700E1	68	$3300.00
Sportavia/Limbach SL1700E	68	$3350.00
Limbach L1700EA1	60	$3450.00
Limbach L1700EA	60	$3400.00
Limbach L2000E01 & EO	80	$4190.00
Limbach L2000EA1	80	$4440.00
Limbach L2000EA	80	$4360.00
Limbach L2000EB (dual carb)	80	$4690.00
Limbach L2000EC1 & EC (pusher)	80	$4330.00
Limbach L2000DD1 (dual mag)	76	$4800.00
Limbach L2000DD2 (dual mag)	76	$4850.00

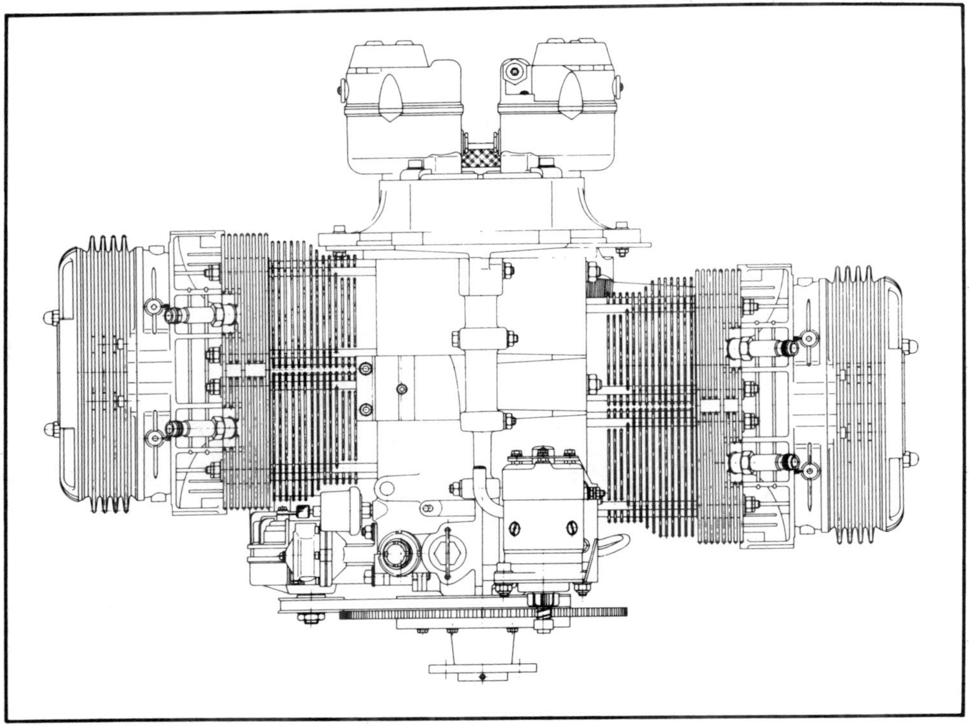

Fig. 6-4. Top view of the Limbach L2000-DD1. (courtesy Limbach Aircraft Corp.)

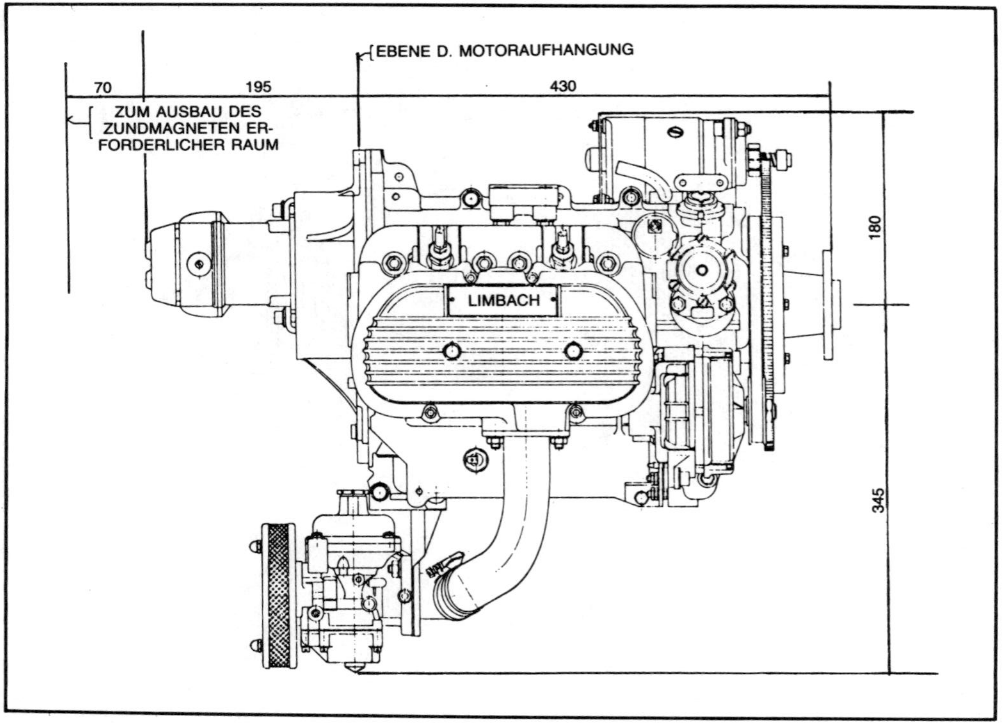

Fig. 6-5. Side elevation of the Limbach L2000-DD1. (courtesy Limbach Aircraft Corp.)

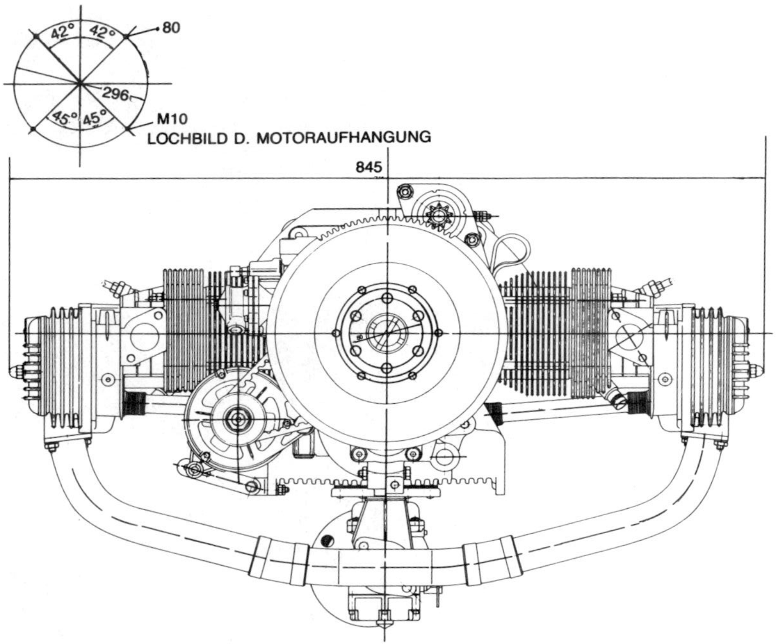

Fig. 6-6. Rear view of the L2000-DD1. The suffix "1" indicates oil system for the German-built Hoffman variable-pitch propeller. (courtesy Limbach Aircraft Corp.)

Fig. 6-7. Neil Bingham of Salt Lake City built this beautiful KR-2, which is powered with the Limbach L2000-EO of 80 hp. (courtesy Neil Bingham)

Note: the ending suffix "1" indicates propeller hub for the German Hoffman Model HO-V62 variable pitch propeller; (all others plus "DD2" have hubs for fixed pitch propellers)

* Delivery to most U.S. locations costs approximately $250.00

For additional information, contact:
Limbach Flugmotoren GmbH & Co.
Kotthausener Str. 5
D-5330 Konigswinter 21,
Sassenberg, West Germany
Phone: 022 44-2322-30-31

Fig. 6-8. New Limbach L550-E, designed for military drones, is a two-cycle, four-cylinder engine of 550cc which produces 40 to 50 hp. (courtesy Limbach Aircraft Corp.)

Limbach Aircraft Engines
P.O. Box 1201,
Tulsa, OK 74101
USA
Phone: (918) 832-9017

Chapter 7

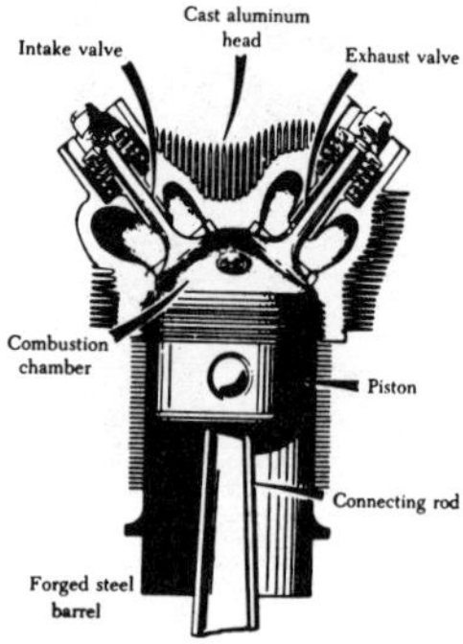

Aircraft Weight and Balance

In college, I had a professor who was fond of saying, "If this discussion is going to get anywhere, let's first define our terms." That certainly applies here; we should define our terms before discussing the procedure for establishing the CG (center of gravity) and loading envelope of the airplane.

Terms

Maximum, or *gross weight*: The maximum weight was fixed by the designer and is listed in the original aircraft specifications. Don't tamper with it. If your machine is amateur-built and licensed in the Experimental category, you may—theoretically, at least—increase the maximum weight if the FAA inspector will okay it, but then you are changing all the original engineering data and stress analysis, and a lot more than a weight/balance problem must be refigured.

Installing a converted automobile engine in an airplane originally designed for an air-cooled aircraft engine usually does impose a weight penalty, but those few extra pounds should be added to airplane's empty weight (subtracted from the useful load) rather than to the maximum or gross weight.

So maximum weight is the all-up total weight of the airplane, including crew, passengers, baggage, fuel—everything that the wings must lift.

Empty weight: The airplane's empty weight includes all operating equipment that has a fixed location and is actually installed in the aircraft. It includes the weight of the airframe, powerplant, required equipment, optional and special equipment, fixed ballast, full engine coolant, hydraulic fluid, and residual fuel in fuel lines, carburetor, and bottom of tanks that is not useable. The FAA says that the engine oil should be drained. If you want to include the engine oil, figure it at 7.5 pounds per gallon and indicate on the weight-balance report whether the aircraft empty weight includes full oil or drained oil.

Useful load: The useful load is the empty weight subtracted from the maximum weight of the airplane. This load consists of the pilot, crew, maximum oil, fuel, passengers, and baggage, unless the oil is otherwise noted.

Weight check: A weight check consists of checking the sum of the weights of all items of useful load against the authorized useful load (maximum weight less empty weight).

Datum: The datum is an imaginary vertical plane from which all horizontal measurements are taken for balance purposes with the aircraft in level flight attitude. The datum is indicated on most FAA Aircraft Specifications. When breaking new ground, any convenient datum may be selected. Once the datum is selected, all moment arms and the location of the permissible CG range must be taken with reference to it. Examples of typical locations of the datum are shown in Fig. 7-1.

Moment arm: The moment arm, or *arm*, is the horizontal distance in inches from the datum to the center of gravity of the machine. If measured aft of the datum, the algebraic sign is plus (+), and minus (−) if measured forward of the datum. Examples of plus and minus arms are shown in Fig. 7-2.

Moment: Moment is the product of a weight multiplied by its arm. The moment of an item about the datum is obtained by multiplying the weight of the item by its horizontal distance from the datum. A typical moment calculation is given in Fig. 7-3.

Center of gravity: The center of gravity (CG) is a point about which the nose-heavy and tail-heavy moments are exactly equal in magnitude. If the aircraft were suspended from there, it would have no tendency to pitch in either direction (nose up or down). The weight of the aircraft may be assumed to be concentrated at its center of gravity (Fig. 7-4).

Empty weight center of gravity: The empty weight CG is the center of gravity of an aircraft in its empty weight condition, and is an essential part of the weight and balance record. Formulas for

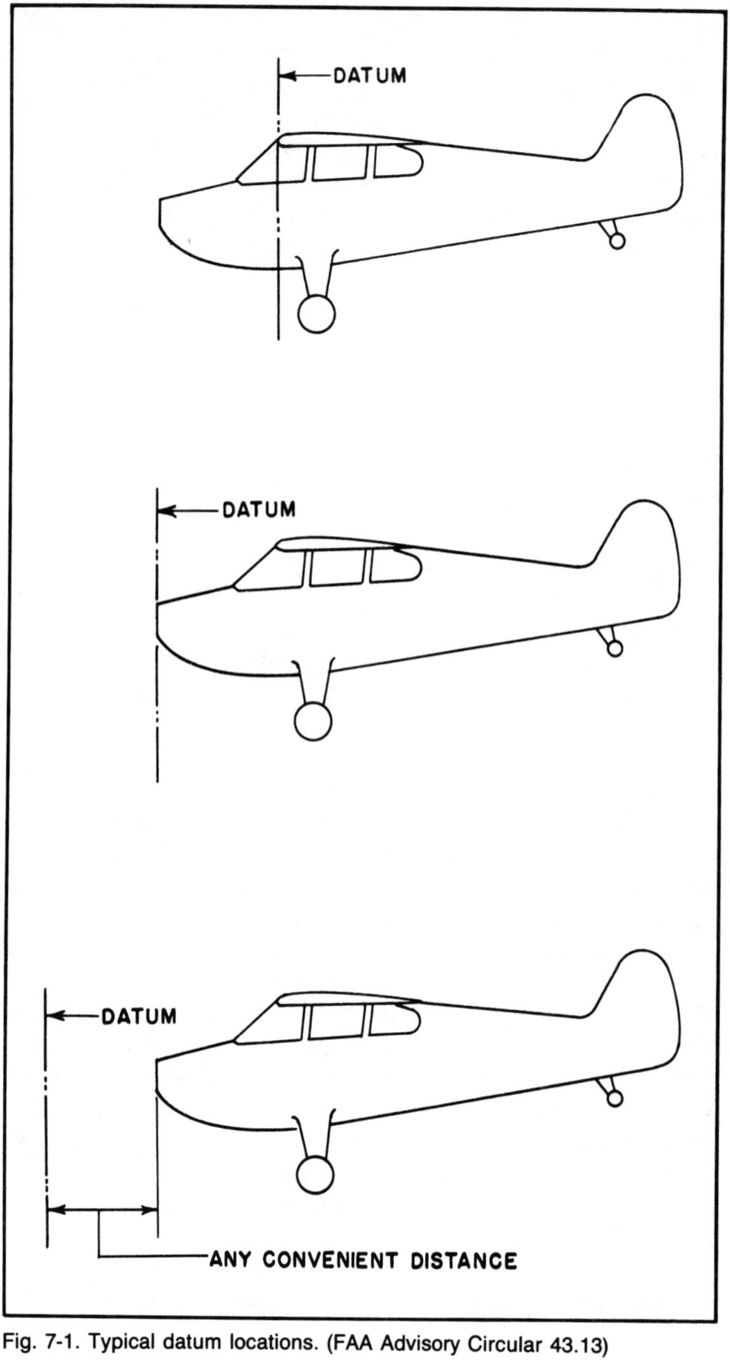

Fig. 7-1. Typical datum locations. (FAA Advisory Circular 43.13)

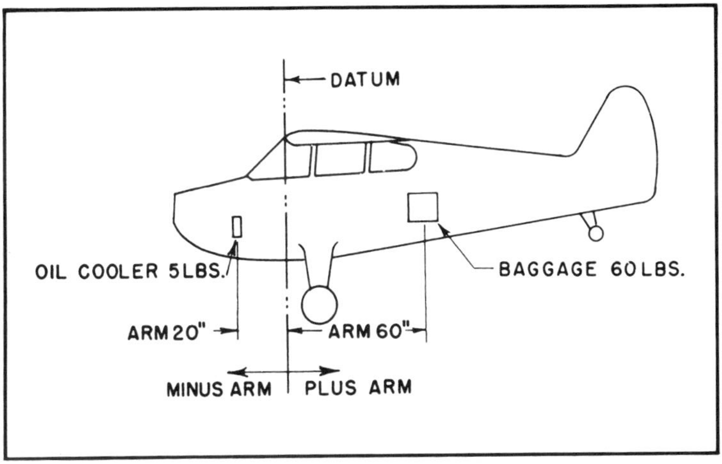

Fig. 7-2. Illustration of arm (or moment arm). (FAA Advisory Circular 43.13)

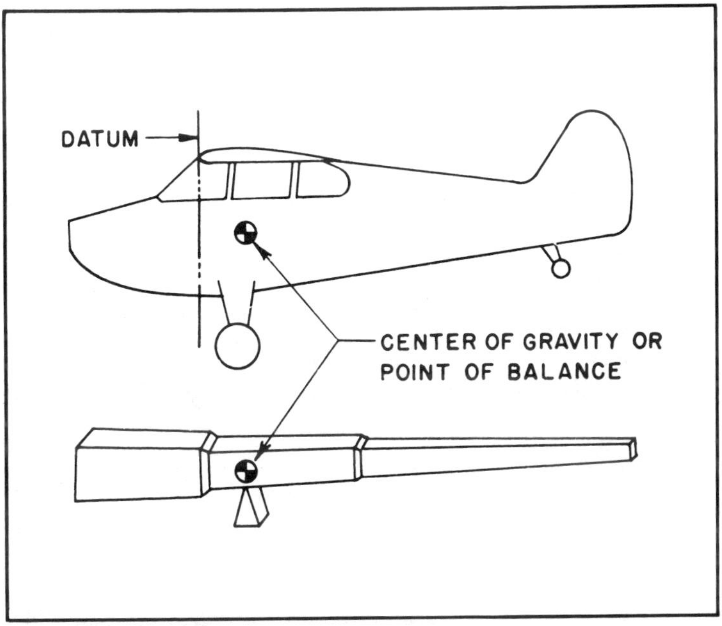

Fig. 7-3. Example of moment computation. The entire aircraft weight may be considered to be concentrated at the center of gravity. Therefore, the moment of the aircraft about the datum is the weight of the aircraft times the horizontal distance between the CG and the datum. Example: If the weight of this airplane is 2,000 pounds, and the arm of the datum to the center of gravity is 16 inches, the moment of the aircraft about the datum is 2,000 × 16, or 32,000 inch/lbs. (FAA Advisory Circular 43.13)

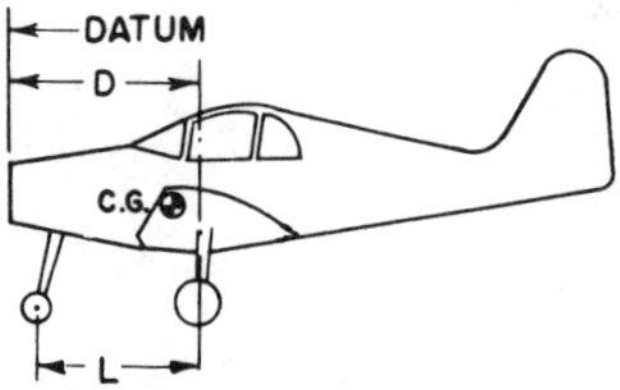

NOSE WHEEL TYPE AIRCRAFT

DATUM LOCATED FORWARD OF THE MAIN WHEELS

$$C.G. = D - \left(\frac{F \times L}{W}\right)$$

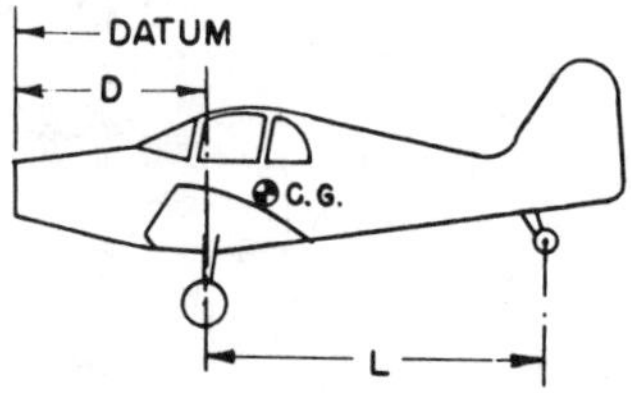

TAIL WHEEL TYPE AIRCRAFT

DATUM LOCATED FORWARD OF THE MAIN WHEELS

$$C.G. = D + \left(\frac{R \times L}{W}\right)$$

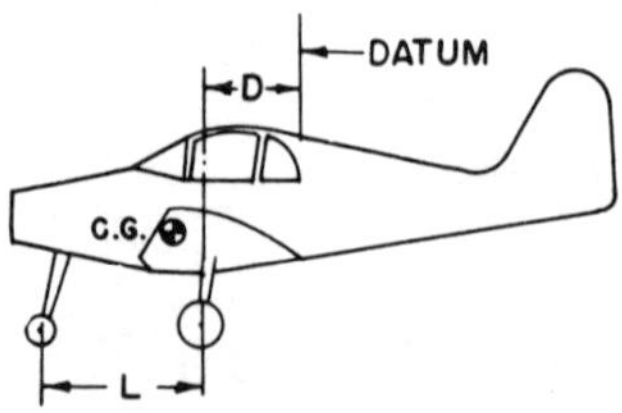

NOSE WHEEL TYPE AIRCRAFT

DATUM LOCATED AFT OF THE MAIN WHEELS

$$C.G. = -\left(D + \frac{F \times L}{W}\right)$$

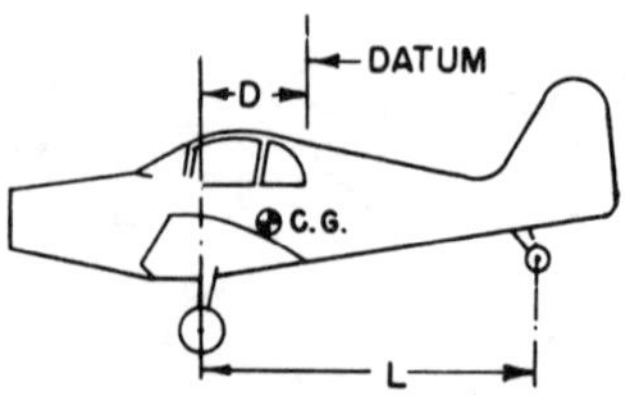

TAIL WHEEL TYPE AIRCRAFT

DATUM LOCATED AFT OF THE MAIN WHEELS

$$C.G. = -D + \left(\frac{R \times L}{W}\right)$$

CG = Distance from datum to center of gravity of the aircraft.
W = The weight of the aircraft at the time of weighing.
D = The horizontal distance measured from the datum to the main wheel weighing point.
L = The horizontal distance measured from the main wheel weighing point to the nose or tail weighing point.
F = The weight at the nose weighing point.
R = The weight at the tail weighing point.

Fig. 7-4. Empty weight center of gravity formulas. (FAA Advisory Circular 43.13)

determining the center of gravity for tail and nosewheel type aircraft are given in Fig. 7-5. Examples of computing the empty weight CG for both types of aircraft are shown in Fig. 7-6.

Empty weight center of gravity range: The empty weight center of gravity range is determined so that the empty weight CG limits will not be exceeded under standard specification loading arrangements. In cases where it is possible to load an aircraft in a manner not covered in the aircraft specifications (i.e., extra tanks, heavier engine, radiator locations, etc.), complete calculations as outlined

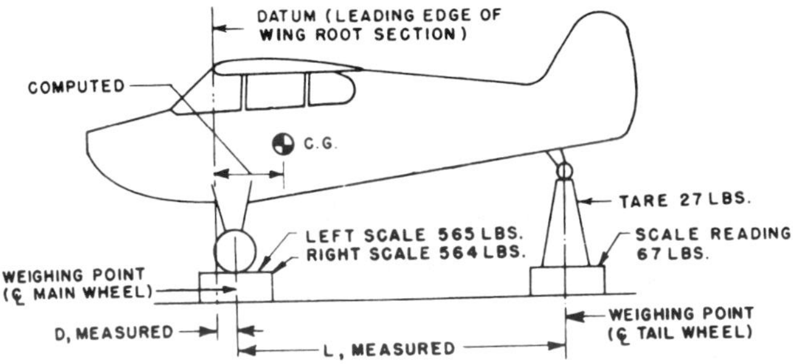

TO FIND: EMPTY WEIGHT AND EMPTY WEIGHT CENTER OF GRAVITY

Datum is the leading edge of the wing (from aircraft specification)
(D) Actual measured horizontal distance from the main wheel weighing point ℄ main wheel) to the Datum---3''
(L) Actual measured horizontal distance from the rear wheel weighing point (℄ rear wheel) to the main wheel weighing point-----------------------------------222''

SOLVING: EMPTY WEIGHT

Weighing Point	Scale Reading #	Tare #	Net Weight #
Right	564	0	564
Left	565	0	565
Rear	67	27	40
Empty Weight (W)			1169

SOLVING: EMPTY WEIGHT CENTER OF GRAVITY

Formula: $C.G. = D + \frac{R \times L}{W} = 3'' + \frac{40 \times 222}{1169} = 3'' + 7.6'' = 10.6''$

Fig. 7-5. Empty weight and empty weight center of gravity for a tailwheel airplane. This case is shown properly entered on a sample weight and balance form, Fig. 7-16. (FAA Advisory Circular 43.13)

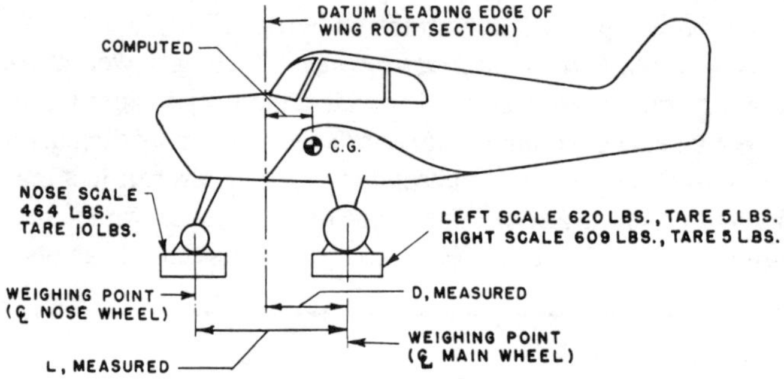

TO FIND: EMPTY WEIGHT AND EMPTY WEIGHT CENTER OF GRAVITY

Datum is the leading edge of the wing (from aircraft specification)
(D) Actual measured horizontal distance from the main wheel weight point (℄ main wheel) to the Datum-- 34.0''
(L) Actual measured horizontal distance from the front wheel weighing point (℄ front wheel) to the main wheel weighing point------------------------------------ 67.8''

SOLVING: EMPTY WEIGHT

Weighing Point	Scale Reading #	Tare #	Net Weight
Right	609	5	604
Left	620	5	615
Front	464	10	454
Empty Weight (W)			1673

SOLVING: EMPTY WEIGHT CENTER OF GRAVITY

Formula: $C.G. = D - \frac{F \times L}{W} = 34'' - \frac{454 \times 67.8}{1673} = 34'' - 18.3'' = 15.7''$

Fig. 7-6. Empty weight and empty weight center of gravity, nosewheel-type airplane. (FAA Advisory Circular 43.13)

under the sub-heading "Loading Conditions and/or Placards" which follows momentarily.

Operating center of gravity range: The operating CG range is the distance between the forward and rearward center of gravity limits indicated by the aircraft specifications. These limits were determined as to the most forward and most rearward loaded CG positions at which the aircraft meets the requirements of the FARs.

The limits are indicated on the specification in either percent of mean aerodynamic chord, or inches from the datum. The CG must be within these limits at all times as illustrated in Fig. 7-7.

Mean aerodynamic chord (MAC): For weight and balance purposes, MAC is used to locate the CG range of the aircraft. The location and dimensions of the MAC will be found in the aircraft specs, flight manual, or the weight and balance report.

Weighing point: If the CG location is determined by weighing, it is necessary to obtain horizontal measurements between the points on the scale at which the aircraft's weight is concentrated. If weighed, using scales under the landing gear tires, a vertical line passing through the centerline of the axle will locate the point on the scale at which the weight is concentrated. This is called the weighing point. Other structural locations capable of supporting the aircraft, such as jack pads on the main spar, may also be used if the airplane weight is resting on the pads. Indicate these points clearly in the weight and balance report when used in lieu of the landing gear. Typical locations of the weighing points are shown in Fig. 7-8.

Full oil: Full oil is the quantity of oil shown in the aircraft's specs as oil capacity. Use full oil as the quantity of oil when making the loaded weight and balance computations.

Tare: Tare is the weight of chocks, blocks, stands, etc., used when weighing the airplane, and which are included in the scale

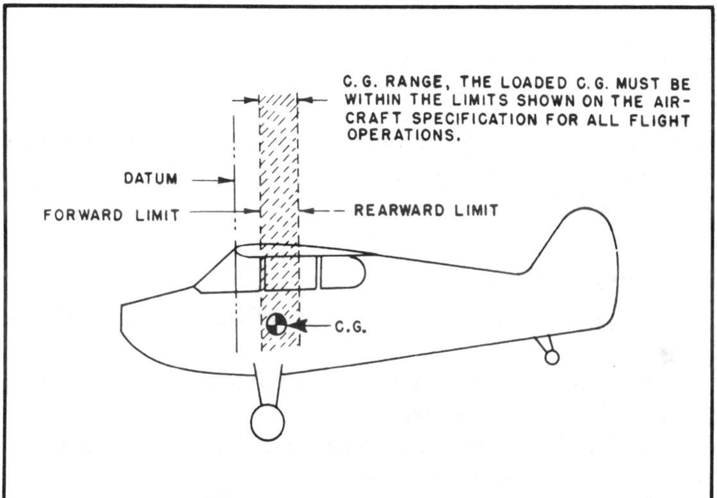

Fig. 7-7. Operating center of gravity range. (FAA Advisory Circular 43.13)

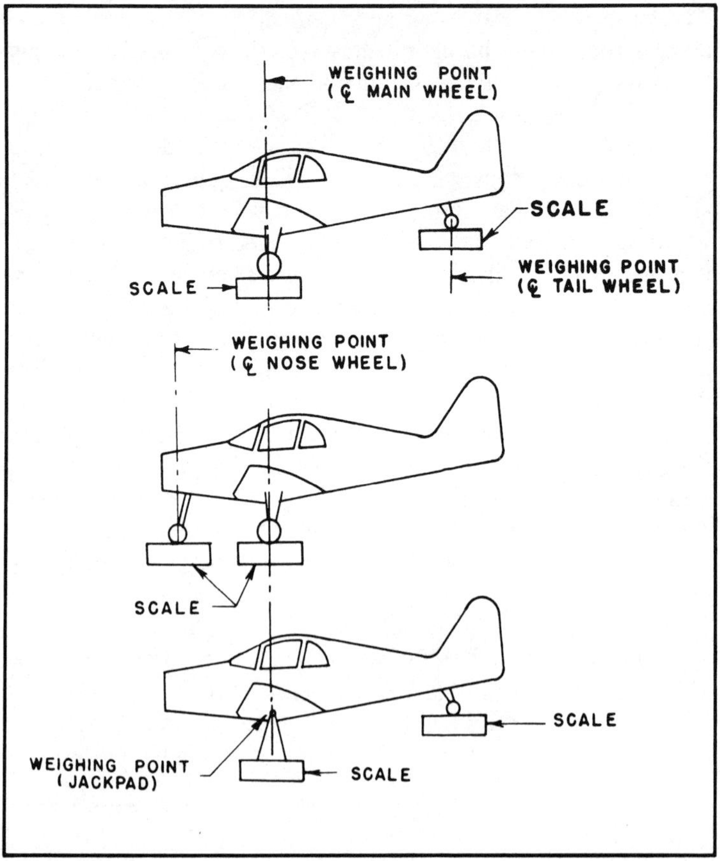

Fig. 7-8. Weighing point centerline. (FAA Advisory Circular 43.13)

readings. Tare is deducted from the scale reading at each respective weighing point where tare is involved to obtain the actual aircraft weight.

Weighing Procedure

A step-by-step procedure for weighing the airplane is as follows:

1. Remove excessive dirt grease, moisture, etc., from the airframe and engine before weighing.
2. Weigh the airplane inside a closed building to prevent error in scale reading due to wind.
3. To determine the center of gravity, place the airplane in a level flight attitude.

4. Have all items of equipment included in the certificated empty weight (if ATCed machine; designed empty weight if homebuilt) installed in the aircraft when weighing. These items will be part of the current weight and balance report.

5. Properly calibrate, zero, and use the scales in accordance with the scales' manufacturer's instructions. The scales and suitable support for the aircraft, if necessary, are usually placed under the wheels of a landplane, the keel of a seaplane float, or the skis of a ski-equipped machine. Other structural locations capable of supporting the aircraft, such as jack pads, may be used. Clearly indicate these points in the weight and balance report.

6. Unless otherwise noted in the aircraft specifications, drain the fuel system until the gauges read "zero" or empty, with the airplane in level flight attitude. The amount of fuel remaining in the tank, lines, and engine is residual fuel, as previously defined, and is to be included in the empty weight.

7. Unless otherwise noted, the oil system should be completely drained with all drain cocks open. Under these conditions the amount of oil remaining in the oil tank, lines and engine, is residual oil.

8. Do not set brakes while taking scale reading.

9. Note the tare when the airplane is removed from the scales.

Weight and Balance Computations

It is necessary, after completing an extensive alteration, total refurbishing of an airplane, or upon completion of a homebuilt, to establish by computation that the proper or authorized weight or CG limits are not exceeded. Paragraph B below explains the significance of the algebraic signs employed.

The aircraft specifications should contain the following information relating to the subject:

1. Center of gravity range.
2. Empty weight CG range when applicable.
3. Leveling means.
4. Datum.
5. Maximum weights.
6. Number of seats and arm.
7. Maximum baggage and arm.
8. Fuel capacity and arm.
9. Oil capacity and arm.
10. Equipment items and arm.

A. Unit Weight for Weight and Balance Purposes:
Gasoline......................6 pounds per U.S. gal.
Lubricating oil................7.5 pounds per U.S. gal.
Crew & passengers.............170 pounds per person.

B. Algebraic Signs: It is important to retain the proper algebraic sign through all balance computations. For the sake of uniformity, visualize the aircraft with its nose to the left. In this position any arm to the left (forward) of the datum is "minus" and any arm to the right (rearward) of the datum is "plus." Any weight added on either side of the datum is plus weight. Any weight removed is minus weight. When multiplying weights by arms, the answer is plus if the signs are alike, and minus if the signs are unalike.

The following combinations are possible:

Weight added forward of the datum:
(+) weight × (−) arm = (−) moment

Weight added to the rear of the datum:
(+) weight × (+) arm = (+) moment

Weight removed forward of the datum:
(−) weight × (−) arm = (+) moment

Weight removed aft of the datum:
(−) weight × (+) arm = (−) moment

The total moment of the aircraft is the algebraic sum of the empty weight moment of the aircraft and all of the individual moments of the weight added and/or subtracted.

Weight and Balance Extreme Conditions

The weight and balance extreme conditions represent the maximum forward and rearward CG position for the airplane. Include in the weight and balance data information showing that the CG of the aircraft (usually in the fully loaded condition) falls between the extreme conditions.

A. Forward Weight and Balance Check. When a forward weight and balance check is made, establish that neither the maximum weight nor the forward CG limit is exceeded as listed in the aircraft specifications. To make this check, the following information is needed.

1. The weights, arms, and moment of the empty airplane.

2. The maximum weights, arms, and moments of the items of useful load which are located ahead of the forward CG limit; and

3. The minimum weights, arms, and moments of the items of useful load which are located aft of the forward CG limit.

A typical example of the computation necessary to make this check, using the above data, is taken from the FAA's Advisory Circular 43.13, Fig. 7-9.

B. Rearward Weight and Balance Check. When a rearward weight and balance check is made, establish that neither the maximum weight nor the rearward CG limit listed in the airplane specification is exceeded. To make this check, the following information is needed:

1. The weight, arms, and moments of the empty airplane.

2. The maximum weights, arms, and moments of the items of useful load which are located aft of the rearward CG limit; and

3. The minimum weights, arms, and moments of the items of useful load which are located ahead of the rearward CG limit.

A typical example of the computation necessary to make this check, using the above data, is shown in Fig. 7-10.

Loading Conditions and/or Placards

If the following items have not been covered in the weight and balance extreme condition checks, or are not covered by suitable placards in the airplane, addition computations are necessary.

These computations should indicate the permissible distribution of fuel, passengers, and baggage which may be carried in the aircraft at any one time without exceeding either the maximum weight or CG range. The conditions to check are:

A. With full fuel, determine the number of passengers and baggage permissible.

B. With maximum passengers, determine the fuel and baggage permissible.

1. Examples of the computations for the above items are given in Figs. 7-11 and 7-12, respectively. The above cases are mainly applicable to lightplanes and homebuilts. Larger transport aircraft have a loading schedule (Fig. 7-13).

Equipment List

A list of the equipment included in the certificated or designed

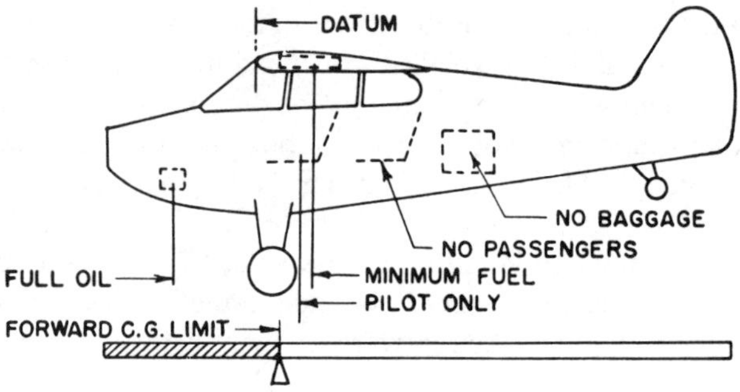

TO CHECK: MOST FORWARD WEIGHT AND BALANCE EXTREME.

GIVEN: Actual empty weight of the airplane-------------- 1169#
Empty weight center of gravity ----------------- +10.6''
*Maximum weight ---------------------------- 2100#
*Forward C.G. limit -------------------------- + 8.5''
*Oil, capacity 9 qts. -------------------------- 17# at - 49
*Pilot in farthest forward seat equipped with controls (unless otherwise placarded) ----------- 170# at + 16''
*Since the fuel tank is located to the rear of the forward C.G. limit, minimum fuel should be included. $\frac{\text{METO HP}}{12} = \frac{165}{12} = 13.75$ gal. x 6# ------- 83# at + 22''

*Information should be obtained from the aircraft specification.

Note: Any items or passengers must be used if they are located ahead of the forward C.G. limit.
Full fuel must be used if the tank is located ahead of the forward C.G. limit.

CHECK OF FORWARD WEIGHT AND BALANCE EXTREME

	Weight (#)	x Arm ('')	= Moment (''#)
Aircraft Empty	+ 1169	+ 10.6	+ 12391
Oil	+ 17	- 49	- 833
Pilot	+ 170	+ 16	+ 2720
Fuel	+ 83	+ 22	+ 1826
Total	+ 1439 (TW)		+ 16104 (TM)

Divide the TM (Total Moment) by the TW (Total Weight) to obtain the forward weight and balance extreme.

$$\frac{TM}{TW} = \frac{16104}{1439} = +\ 11.2''$$

Since the forward C.G. limit and the maximum weight are not exceeded, the forward weight and balance extreme condition is satisfactory.

Fig. 7-9. Example of check of most forward weight and balance extreme. (FAA Advisory Circular 43.13)

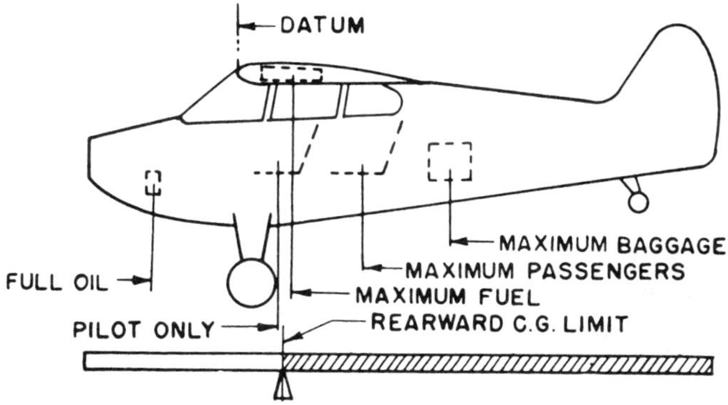

TO CHECK: MOST REARWARD WEIGHT AND BALANCE EXTREME.

GIVEN: Actual empty weight of the airplane --------- 1169#
Empty weight center of gravity ------------- 10.6''
*Maximum weight ------------------------- 2100#
*Rearward C.G. limit --------------------- 21.9''
*Oil capacity 9 qts. ----------------------- 17# at - 49''
*Baggage, placarded do not exceed 100 lbs. --- 100# at + 75.5''
*Two passengers in rear seats, 170 x 2 ------ 340# at + 48''
*Pilot in most rearward seat equipped with controls (unless otherwise placarded) ------- 170# at + 16''
*Since the fuel tank is located aft of the rearward C.G. limit full fuel must be used --- 240# at + 22''

* Information should be obtained from the aircraft specification.

Note: If fuel tanks are located ahead of the rearward C.G. limit minimum fuel should be used.

CHECK OF REARWARD WEIGHT AND BALANCE EXTREME

	Weight (#)	x Arm ('')	= Moment (''#)
Aircraft empty	+ 1169	+ 10.6	+ 12391
Oil	+ 17	- 49	- 833
Pilot (1)	+ 170	+ 16	+ 2720
Passengers (2)	+ 340	+ 48	+ 16320
Fuel (40 gals.)	+ 240	+ 22	+ 5280
Baggage	+ 100	+ 75.5	+ 7550
Total	+ 2036 (TW)		+ 43428 (TM)

Divide the TM (Total Moment) by the TW (Total Weight) to obtain the rearward weight and balance extreme.

$$\frac{TM}{TW} = \frac{43428}{2036} = +\ 21.3''$$

Since the rearward C.G. limit and the maximum weight are not exceeded, the rearward weight and balance extreme condition is satisfactory.

Fig. 7-10. Example of check of most rearward weight and balance extreme. (FAA Advisory Circular 43.13)

EXAMPLE OF THE DETERMINATION OF THE NUMBER OF PASSENGERS AND BAGGAGE PERMISSIBLE WITH FULL FUEL

GIVEN:

Actual empty weight of the aircraft ------------------	1169#
Empty weight center of gravity ----------------------	10.6"
Maximum weight ------------------------------------	2100#
Datum is leading edge of the wing	
Forward center of gravity limit ---------------------	8.5"
Rearward center of gravity limit ---------------------	21.9"
Oil capacity, 9 qts.; show full capacity ----------------	17# at -49"
Baggage, maximum ---------------------------------	100# at +75.5"
Two passengers in rear seat, 170# x 2 ----------------	340# at +48"
Pilot in most rearward seat equipped with controls (unless otherwise placarded) ----------------	170# at +16"
Full fuel, 40 gals. x 6# ----------------------------	240# at +22"

	Weight(#)	x Arm(")	= Moment("#)
Aircraft empty	+ 1169	+ 10.6	+ 12391
Oil	+ 17	- 49	- 833
Full fuel	+ 240	+ 22	+ 5280
Passengers, 2 rear	+ 340 *	+ 48	+ 16320
Pilot	+ 170	+ 16	+ 2720
Baggage	+ 100	+ 75.5	+ 7550
Total	+ 2036 (TW)		+ 43428 (TM)

Divide the TM (total moment) by the TW (total weight) to obtain the loaded center of gravity.

$$\frac{TM}{TW} = \frac{43428}{2036} = +21.3''$$

The above computations show that with full fuel, 100 pounds of baggage and two passengers in the rear seat may be carried in this aircraft without exceeding either the maximum weight or the approved C.G. range.

This condition may be entered in the loading schedule as follows:

GALLONS OF FUEL	NUMBER OF PASSENGERS	POUNDS OF BAGGAGE
Full	2 Rear	100

* Only two passengers are listed to prevent the maximum weight of 2100 lbs. from being exceeded.

Fig. 7-11. Loading conditions: determination of the number of passengers and baggage permissible with full fuel. (FAA Advisory Circular 43.13)

empty weight should be found in the airplane flight manual or on the weight and balance report. Enter in the weight and balance report all the required, optional, and special equipment installed in the airplane at the time of weighing and/or subsequent equipment changes.

A. Required equipment items are items so listed in the pertinent aircraft specification.

B. Optional equipment items are so listed in the specs and are, of course, installed in the airplane at the option of the owner.

C. Special equipment includes any item not corresponding exactly to the descriptive information in the specs. This includes radios, position lights, carpets, instruments, ashtrays, etc.

D. Required and optional equipment may be shown on the equipment list making reference to the pertinent item number listed in the applicable specification only when they are identical to that number item with reference to description, weight, and arm given in the specs. Show all special equipment items by making reference to the item by name, make, model, weight, and arm. When

EXAMPLE OF THE DETERMINATION OF THE POUNDS OF FUEL AND BAGGAGE PERMISSIBLE WITH MAXIMUM PASSENGERS

	Weight (#)	x Arm (")	= Moment ("#)
Aircraft empty	+ 1169	+ 10.6	+ 12391
Oil	+ 17	- 49	- 833
Pilot	+ 170	+ 16	+ 2720
Passenger, 1 front	+ 170	+ 16	+ 2720
Passengers, 2 rear	+ 340	+ 48	+ 16320
Fuel (39 gals.)	+ 234	+ 22	+ 5148
Baggage	---	---	---
Total	+ 2100		+ 38466

Divide the TM (total moment) by the TW (total weight) to obtain the loaded center of gravity.

$$\frac{TM}{TW} = \frac{38466}{2100} = +18.3''$$

The above computations show that with the maximum number of passengers, 39 gallons of fuel and zero pounds of baggage may be carried in this aircraft without exceeding either the maximum weight or the approved C.G. range.

This condition may be entered in the loading schedule as follows:

GALLONS OF FUEL	NUMBER OF PASSENGERS	POUNDS OF BAGGAGE
* FULL	* 2 rear	* 100
39	1(F) 2(R)	None

Fig. 7-12. Loading conditions: determination of the fuel and baggage permissible with maximum passengers.
*Conditions as entered from Fig. 7-11.
(F) Front seat; (R) Rear seat.

the arm for such an item is not available, determine by actual measurement.

1. Equipment changes require an entry on the equipment list indicating items added, removed, or relocated, with the date accomplished, and the craftsman's name, along with certificate number.

Examples of items so affected include the installation of additional fuel tanks, seats, and baggage compartments. Figure 7-14 illustrates the effect on balance when equipment items are added within the acceptable CG limits and fore and aft of the established CG limits. Moment computations for typical equipment changes are given in Fig. 7-15, and are also included in the sample weight and balance sheet in Fig. 7-17.

EXAMPLE OF THE DETERMINATION OF THE FUEL AND THE NUMBER AND LOCATION OF PASSENGERS PERMISSIBLE WITH MAXIMUM BAGGAGE

	Weight (#)	x Arm ('')	= Moment (''#)
Aircraft empty	+ 1169	+ 10.6	+ 12391
Oil	+ 17	- 49	- 833
Pilot	+ 170	+ 16	+ 2720
Passenger (1) rear	+ 170	+ 48	+ 8160
Passenger (1) front	+ 170	+ 16	+ 2720
Fuel (40 gals.)	+ 240	+ 22	+ 5280
Baggage	+ 100	+ 75.5	+ 7550
Total	+ 2036		+ 37988

Divide the TM (total moment) by the TW (total weight) to obtain the loaded center of gravity.

$$\frac{TM}{TW} = \frac{37988}{2036} = +18.7$$

The above computations show that with maximum baggage, full fuel and 2 passengers (1 in the front seat and 1 in the rear seat) may be carried in this aircraft without exceeding either the maximum weight or the approved C.G. range.

This condition may be entered in the loading schedule as follows:

GALLONS OF FUEL	NUMBER OF PASSENGERS	POUNDS OF BAGGAGE
* Full	* 2 Rear	*100
** 39	*1(F) 2(R)	**None
Full	1(F) 1(R)	Full

Fig. 7-13. Loading conditions: determination of the fuel and number and location of passengers permissible with maximum baggage.
* Conditions are entered from Fig. 7-11.
** Conditions as entered from Fig. 7-12.
(F) Front seat; (R) Rear seat.

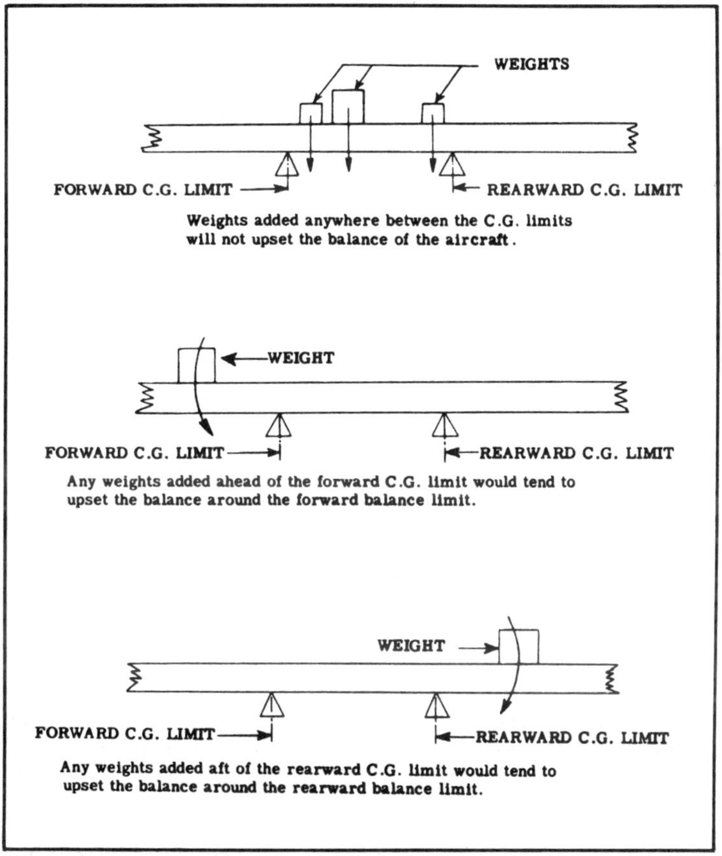

Fig. 7-14. Effects of additional equipment items on balance. (FAA Advisory Circular 43.13)

Sample Weight and Balance Report

Suggested methods of tabulating the various data and computations for determining the CG, both in the empty weight condition and fully loaded, are given in Figs. 7-16 and 7-17 respectively, and show the FAA's suggested means of recording this information. The data presented in Figs. 7-16 and 7-17 have previously been computed in Figs. 7-9 and 7-10 for the extreme load conditions, and Fig. 7-15 for equipment change and represent suggested means of recording this information.

Installation of Ballast

Ballast is sometimes permanently installed for CG balance pur-

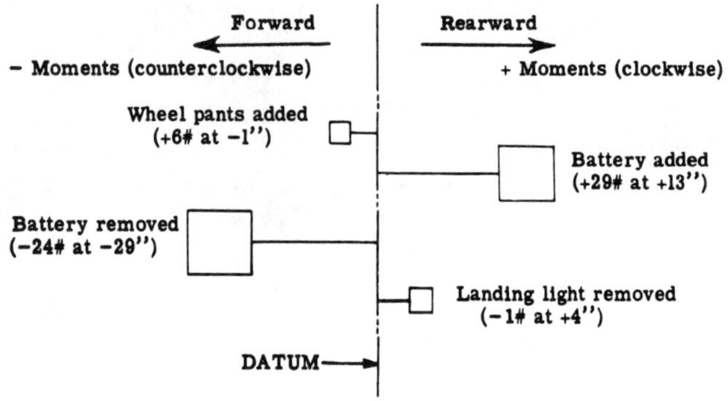

	Weight (#)	x Arm (")	= Moment ("#)
ADDED			
Item 204 wheel pants	+ 6	- 1	- 6
Item 302b Battery	+ 29	+13	+377
REMOVED			
Item 302a Battery	- 24	-29	+696
Item 303 Landing light	- 1	+ 4	- 4

Fig. 7-15. Example of moment and weight changes resulting from equipment changes. This condition is shown properly entered on a sample weight and balance report on Fig. 7-17 under "Equipment Change." (FAA Advisory Circular 43.13)

poses as a result of installation or removal of equipment and is not used to correct a nose-up or nose-down tendency of an airplane. It is usually located as far forward or as far aft as possible in order to bring the CG position within acceptable limits with a minimum of weight increase. Permanent ballast is often in the form of lead plate wrapped around and/or bolted to the fuselage primary structure (tailpost, longerons, or bulkhead). Permanent ballast invariably constitutes a concentrated load; therefore, the strength of the local structure and the attachment of the ballast to it should be investigated for the design loading conditions applying to that airplane. The ballast must be placarded "Permanent Ballast—Do Not Remove."

It is not desirable to install permanent ballast by pouring melted lead into the tailpost or longerons due to difficulties that may be encountered in subsequent welding repair operations.

It should be noted that the installation of permanent ballast results in an increase in aircraft empty weight (see Fig. 7-18 for ballast computation). When disposable ballast is carried, the local strength of the compartment in which the ballast is carried and the effect of the ballast on aircraft balance and weight should be investigated.

Loading Schedule

The loading schedule should be kept with the aircraft and usually forms part of the airplane flight manual. It includes instruc-

MAKE MA-700 MODEL A SERIAL # 0000 REGISTRATION # N1234
DATUM IS leading edge of wing.

COMPUTE AS FOLLOWS IF AIRCRAFT WEIGHED

1. Leveling means: level top longeron between front and rear seats.
2. Main wheel weighing point is located (_____"FORWARD)(+ 3 " AFT) of datum.
3. Actual measured distance from the main weight point centerline to the tail (or nose) point centerline 222 ".
4. Oil over and above "ZERO" tank reading = (a. — Gals.)(b. — Lbs.)(c. — In.)

ACTUAL EMPTY WEIGHT

Weight Point	Scale Reading -	Tare =	Net Weight
5. Right	564	0	564
6. Left	565	0	565
7. Tail	67	27	40
8. Nose	—	—	—
9. Total Net Weight			1169

CENTER OF GRAVITY AS WEIGHED

10. C.G. relative to main wheel weighing point:
(a) Tail wheel airc. (item 3, 222) X (Item 7, 40) / (Item 9, 1169) = + 7.6 = C.G.

(b) Nose wheel airc. (Item 3 —) X (Item 8 —) / (Item 9 —) = _____ = C.G.

11. C.G. relative to datum:
(a) Tail wheel airc. (Item 10a, + 7.6) added to (Item 2, + 3) = + 10.6" = C.G.
(b) Nose wheel airc. (Item 10b, ____) added to (Item 2, ____) = _____ = C.G.

COMPUTE IF AIRCRAFT WEIGHED WITH OIL (Item 4)

	Weight X	Arm =	Moment
Aircraft	(9)	(11)	
Less Oil	(4b)	(4c)	
Empty Totals	(a)		(b)

12. (b) / (a) ______ = (c) ______" = Empty weight C.G.

REPAIR AGENCY __________ (Name) __________ (Number) DATE __________

Fig. 7-16. Sample weight and balance report to determine empty weight center of gravity. (FAA Advisory Circular 43.13)

tions on the proper load distribution, such as the filling of fuel and oil tanks, passenger seating, restrictions on passenger movement, distribution of cargo, etc.

A. Other means of determining safe loading conditions, such as the use of a graphical index, "load adjuster," etc., are acceptable, and may be used in lieu of the information presented above under the subheading "Loading Conditions and/or Placards."

EQUIPMENT LIST

*Required or Optional Item Numbers as Shown in Aircraft Specification						
1	2	101	102	103	104	105
106	201	202	203	301	302(a)	303
401(a)	402	—	—	—	—	—

Special Equipment				
Item	Make	Model	Weight	Arm
3 Flares 1½ Min.	XYZ	03	25#	150"
Enter above those items included in the empty weight.				

WEIGHT AND BALANCE EXTREME CONDITIONS

Approved fwd limit 8.5 "   Approved max. weight 2100#   Approved aft limit 21.9"

Item	FORWARD CHECK Weight	X Arm	= Moment	REARWARD CHECK Weight	X Arm	= Moment
Airc. Empty	+ 1169 (9 or 12a)	+ 10.6 (11 or 12c)	+ 12391	+ 1169 (9 or 12a)	+ 10.6 (11 or 12c)	+ 12391
Oil	+ 17	- 49	- 833	+ 17	- 49	- 833
Pilot	+ 170	+ 16	+ 2720	+ 170	+ 16	+ 2720
Fuel	+ 83	+ 22	+ 1826	+ 240	+ 22	+ 5280
Passenger(s)				+ 340	+ 48	+ 16320
Baggage				+ 100	+ 75.5	+ 7550
TOTAL	+ 1439 = TW		+ 16104 = TM	+ 2036 = TW		+ 43428 = TM

$\frac{TM}{TW} = \frac{16104}{1439} = +11.2" =$ Most Forward C.G. location

$\frac{TM}{TW} = \frac{43428}{2036} = +21.3" =$ Most rearward C.G. location

LOADING SCHEDULE

Gallons of Fuel	Number of Passengers	Pounds of Baggage
40	2(R)	100
The above includes pilot and capacity oil.		

EQUIPMENT CHANGE

Computing New C.G.			
Item, Make, and Model*	Weight X	Arm =	Moment
Airc. Empty	+ 1169 (9 or 12a)	+ 10.6 (11 or 12c)	+ 12391
204 added	+ 6	- 1	- 6
302(b) added	+ 29	+ 13	+ 377
302(a) removed	- 24	- 29	+ 696
303 removed	- 1	+ 4	- 4
NET TOTALS	- 1179 = NW		+ 13454 = NM

$\frac{NM}{NW} = \frac{13454}{1179} = + 11.4" =$ New C.G.

*ITEM NUMBERS WHEN LISTED IN PERTINENT AIRCRAFT SPECIFICATION MAY BE USED IN LIEU OF "ITEM, MAKE, AND MODEL".

PREPARED BY __________ DATE __________

Fig. 7-17. Sample weight and balance report including an equipment change for aircraft fully loaded. (FAA Advisory Circular 43.13)

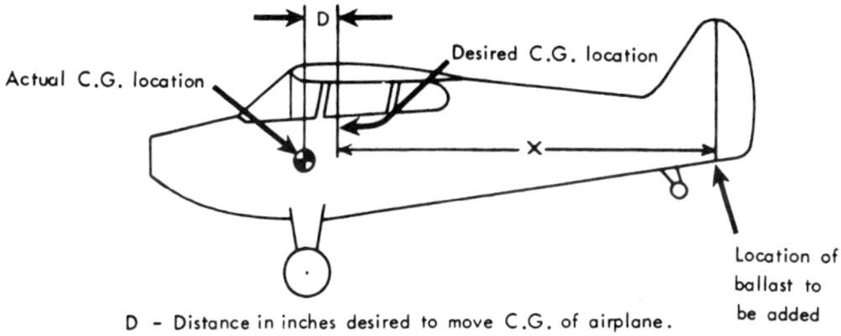

D - Distance in inches desired to move C.G. of airplane.

W - Weight of airplane as loaded.

X - Distance in inches from point where ballast is to be installed, to the desired location of the new C.G.

B - Weight of ballast required in pounds.

$$B = \frac{D \times W}{X}$$

Compute the new C.G. of the aircraft with ballast installed.

NOTE: If greater accuracy is desired, repeat the entire formula using the NEW aircraft weight and the NEW C.G. in the second operation.

Fig. 7-18. Permanent ballast computation formula. (FAA Advisory Circular 43.13)

B. Compute a separate loading condition when the aircraft is to be loaded in other than the specified conditions shown in the loading schedule.

Appendix

Appendix A

Engine Nomenclature

For those readers who may be a bit hazy on the operation principle of the four-stroke cycle gasoline engine, it is explained here, along with piston displacement and compression ratio. After more than 80 years of experimentation and development, the reciprocating engines that power our airplanes and most automobiles operate on the same principle as the one which flew the world's first successful airplane in 1903. These powerplants employ the four-stroke principle—intake, compression, power, and exhaust—which is technically the Otto Cycle, named for Nickolas Otto, who perfected it in 1876.

The Otto Cycle

In these engines, the crankshaft revolves twice for each cycle of four strokes, each piston making two upward strokes away from the crankshaft, and two downward strokes toward the crankshaft. Thus, each cylinder "works" every other revolution of the crankshaft, while the piston coasts through on the intervening revolution and pushes out the burned gases produced by the working stroke of the previous revolution.

A complete cycle is as follows:

As the intake stroke begins, the piston is near top dead center (TDC) and the intake valve is opened by its camshaft lobe. The piston travels downward, sucking the fuel/air mixture—which has been mixed and atomized coming through the carburetor—into the combustion chamber of the upper cylinder by way of the intake manifold. When the piston reaches the end of the intake stroke, the camshaft allows the spring-loaded intake valve to close and the compression stroke begins as the piston now moves upward toward the top of the cylinder. A split-second before the piston reaches TDC, the ignition system sends an electrical charge through the spark plug, igniting the compressed fuel/air mixture. This controlled explosion sends the piston downward on its power stroke. Then, when the piston reaches the bottom of the cylinder, the camshaft opens the exhaust valve, and the piston, coasting through on the exhaust stroke, purges the cylinder of the burned gas. Now the intake valve opens again and the cycle is repeated (Figs. A-1, A-2).

As mentioned, the valves are actuated by the camshaft lobes. The camshaft is geared to revolve at one-half crankshaft speed because the intake and exhaust valves open and close only once for two revolutions of the crankshaft.

Valve timing differs slightly with different engines. On most, the intake valve is opened before the exhaust valve is completely closed. This makes for better expulsion of the burned gases, and is called *valve overlap*.

The spark from the ignition system is nearly always supplied to the combustion chamber before the piston reaches TDC. This is called *spark advance*. For example, a 15-degree spark advance means that the plugs fire when the piston is still 15 degrees below TDC and completion of the compression stroke.

Gasoline knock value is given in octane ratings because it is determined by comparing the knock value of a given fuel with the knock value of a reference fuel mixture of iso octane and heptane.

Piston Displacement

When other factors remain equal, the greater the piston displacement, the greater the maximum horsepower an engine will be capable of developing. There's an ancient saying among engine men: "There's no substitute for cubic inches."

When a piston moves from bottom dead center to top dead center, it displaces a specific volume. The volume displaced by the piston is known as *piston displacement* and is expressed in cubic inches for airplane engines in America (Detroit likes to express piston displacement in liters these days). The piston displacement of one cylinder may be obtained by multiplying the area of the cross section of the cylinder by the total distance the piston moves in

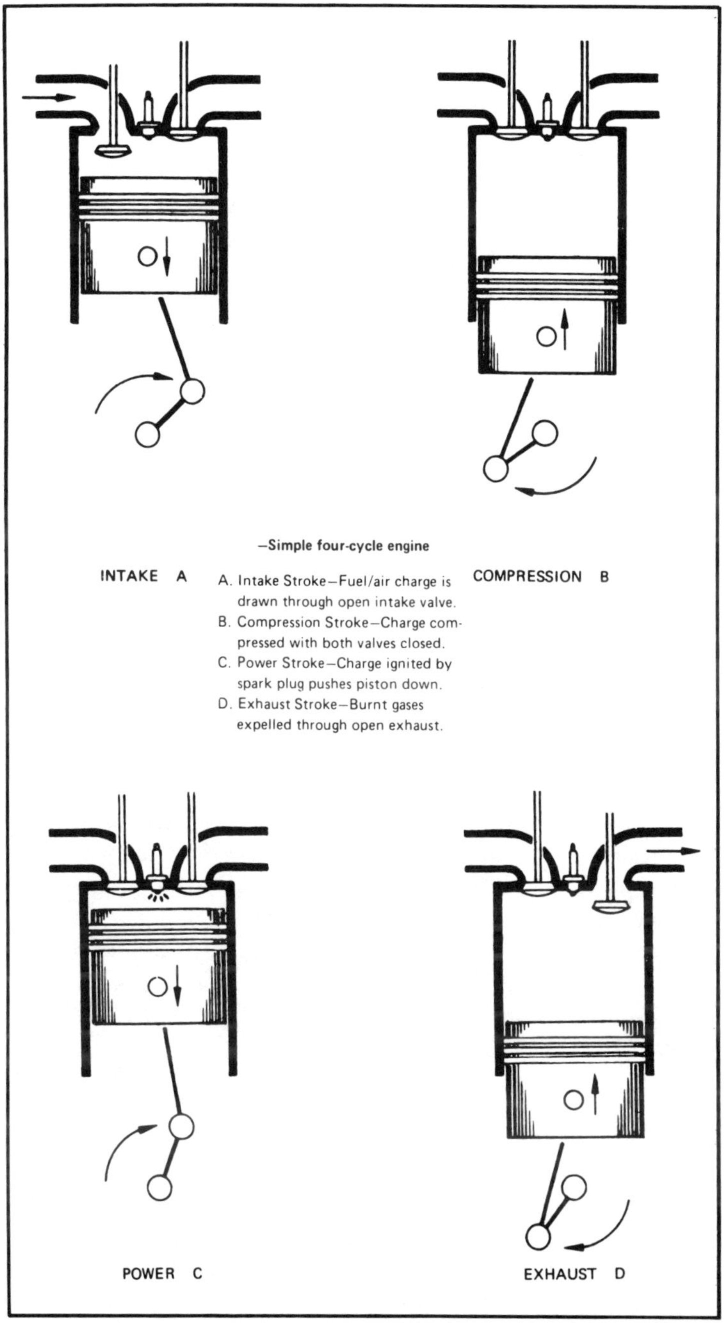

Fig. A-1. The Otto Cycle principle. (courtesy HAPI Engines, Inc.)

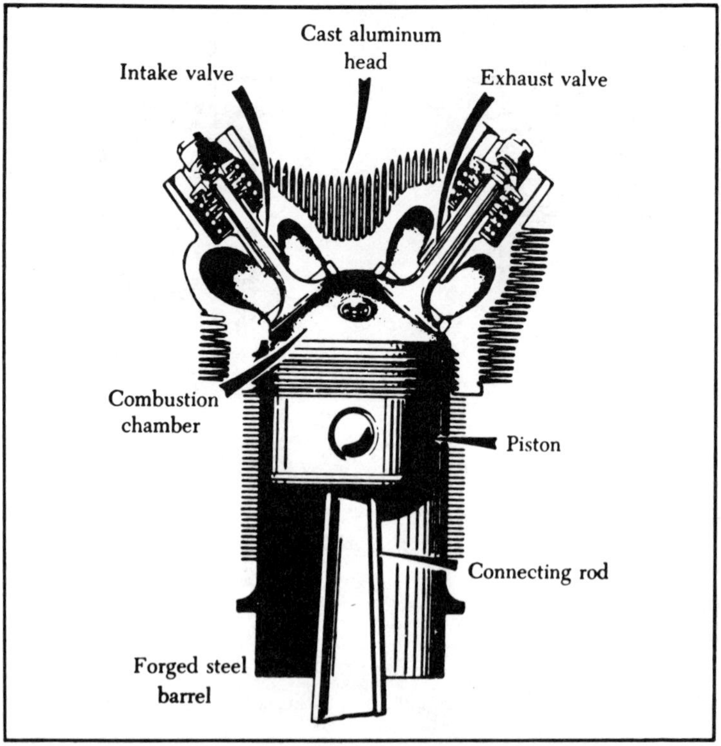

Fig. A-2. Typical air-cooled engine cylinder assembly. (FAA AC 65.12)

the cylinder in one stroke. For multi-cylinder engines this product is multiplied by the number of cylinders to get the total piston displacement of the engine.

Since the volume (V) of a geometric cylinder equals the area (A) of the base multiplied by the altitude (H), it is expressed mathematically as: $V = AH$.

For our purposes, the area of the base is the area of the cross section of the cylinder or of the piston top, and the area of a circle is: $A = 1/4\pi D^2$ where π equals 3.1416, and D is the diameter.

Example:	Given:	Piston diameter	= 5.5 in.
		Stroke	= 5.5 in.
		No. of cylinders	= 7
	Find:	A (area of piston head)	
		V (volume; i.e., displacement of a single piston)	
	Total	V (total volume of all the pistons)	

Substituting these values in the equation:

$$\begin{aligned} A &= 1/4\pi D^2 \\ &= 1/4 \times 3.1416 \times 5.5 \times 5.5 \\ &= 0.7854 \times 30.25 \\ &= 23.758 \text{ sq/in} \end{aligned}$$

The altitude in this case equals the length of the stroke. The volume of the cylinder (displacement of one piston) is therefore the area multiplied by the length of the stroke.

Substituting in the equation:

$$\begin{aligned} V &= AH \\ &= 23.758 \times 5.5 \\ &= 130.669 \text{ cu/in} \end{aligned}$$

This volume multiplied by the number of cylinders will give the total displacement of the engine:

$$130.669 \times 7 = 914.683 \text{ cu/in}$$

The answer is rounded to the nearest 10 inches. Therefore, 914.683 cu/in would be identified as a 900 cubic-inch engine.

Compression Ratio

All internal combustion engines must compress the fuel/air mixture to receive a reasonable amount of work from each power stroke. The fuel/air charge in the cylinder can be compared to a coil spring: The more it is compressed, the more work it is potentially capable of doing.

The compression ratio of an engine (Fig. A-3) is a comparison of the volume of space in a cylinder when the piston is at the bottom of the stroke to the volume of space when the piston is at the top of the stroke. This comparison is expressed as a ratio, hence the term *compression ratio*. Compression ratio is a controlling factor in the maximum horsepower developed by an engine, but it is limited by fuel grades, and the high engine speeds and manifold pressures required for takeoff.

For example, if there are 140 cu/in of space in the cylinder when the piston is at the bottom and there are 20 cu/in of space when the piston is at the top of the stroke, the compression would be

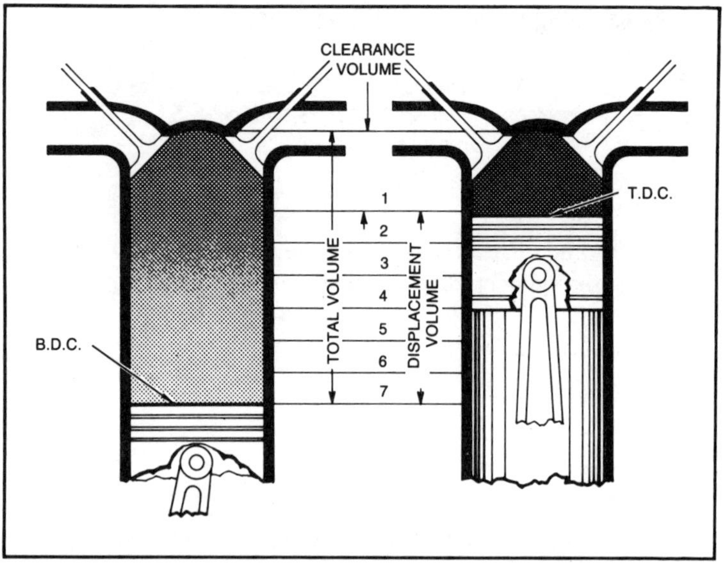

Fig. A-3. Displacement and compression ratio. (FAA AC 65.12)

140:20 Expressed in fraction form, it would be 140/20, or 7 to 1, usually represented as 7:1.

To grasp more thoroughly the limitation placed on compression ratios, manifold pressure and its effect on compression pressures should be understood. *Manifold pressure* is the average absolute pressure of the air or fuel/air charge in the intake manifold and is measured in units of inches of mercury (Hg). Manifold pressure is dependent on engine speed (throttle setting) and supercharging. The engine-driven internal supercharger (blower) and the external exhaust-driven supercharger (turbo) are actually centrifugal-type air compressors. The operation of these superchargers increases the weight of the charge entering the cylinder. When either or both are used with an aircraft engine, the manifold pressure may be considerably higher than the pressure of the outside atmosphere. The advantage of this condition is that a greater amount of fuel/air charge is forced into a given cylinder volume, and greater power results.

Compression ratio and manifold pressure determine the pressure in the cylinder in that portion of the operating cycle when both valves are closed. The pressure of the charge before compression is determined by manifold pressure, while the pressure at the height of compression (just prior to ignition) is determined by manifold pressure times the compression ratio. For example, if an engine were operating at a manifold pressure of 30 inches Hg with a compression ratio of 7:1, the pressure at the instant before ignition would be approximately 210 inches Hg. However, at a manifold pressure of 60 inches Hg the pressure would be 420 inches Hg.

Without going into great detail, it has been shown that the compression event magnifies the effect of varying the manifold pressure, and the magnitude of both affects the pressure of the fuel charge just before the instant of ignition. If the pressure at this time becomes too high, premature ignition, or knock, will occur and produce overheating.

One of the reasons for using engines with high compression ratios is to obtain long-range fuel economy; that is, to convert more heat energy into useful work than is done in engines of low compression ratio. Since more heat of the charge is converted into useful work, less heat is absorbed by the cylinder walls. This factor promotes cooler engine operation, which, in turn, increases thermal efficiency.

Here again, a compromise is needed between the demand for fuel economy and the demand for maximum horsepower without knocking. For years, tetraethyl lead (TEL) has been added to gasoline to allow high compression without knocking. "Regular" automotive gasoline has a high TEL content and a maximum octane rating below 90. It may be used in most aircraft engines rated for 80/87 fuel, but it will eventually disappear.

Automotive unleaded, which is also around 87 octane, represents salvation for all those 80-octane air-cooled ATCed airplane engines out there. The so-called low-lead 100 aviation fuel (blue) burns valves, fouls spark plugs, and contaminates oil in these engines. However, LL100 blue does work okay in the VW conversions; Revmaster says use it, while HAPI engines are happy with automotive unleaded.

Actually, LL100 blue is "low lead" only in relation to 100/130 green. The actual figures for aviation gasoline are: 1/2 milliliter TEL per gallon in 80/87 red; 1.2 to 2 milliliters per gallon in LL100 blue, and 3 to 4 milliliters per gallon in 100/130 green.

One side note on 100 blue: It is more destructive of aircraft finishes than other av-fuels. Cessna gives this as its main reason for switching to polyurethane paints in mid-1977.

Propeller Nomenclature

Propellers are identified according to the length and pitch; e.g., 52 × 44, which means a propeller measuring 52 inches from tip to tip, possessing a pitch of 44 inches. Pitch is the distance the propeller would move forward in one revolution if it were boring into a solid and suffered no slippage. But there *is* slippage, and a prop with a 44-inch pitch does not move forward (and pull the airplane with it) 44 inches in each revolution. This is the only measure we have, however, and it has to do.

The standard place to measure pitch is at 75 percent of the distance from the center of the hub toward the tip. The blades' angle decreases toward the tips, because in rotation, the tips achieve the highest speed. Near the hub, even at high rpm, the blades turn too slowly to produce significant thrust, although the tips may be approaching the speed of sound. So the outer one-third of the prop blades produces most of the thrust, and this outer one-third should not get too close to the speed of sound because then it begins to produce shock waves (noise), but little thrust.

The propeller tips need not actually reach the speed of sound—approximately 1,100 feet per second—before shock waves are generated, because the propeller is an airfoil, in cross section shaped like a wing. It produces "lift" just as the wing does, except in a different direction. It is a subsonic airfoil, and the airflow over its curved surface speeds up just as it does over a wing. Therefore, the airflow over a portion of the blades may actually go supersonic while tip velocities are less than 1,200 fps. That is why a maximum tip speed of about 800 fps is recommended for good efficiency.

The following formula will provide you with the maximum allowable rpm to hold tip speeds below 900 fps:

$$\frac{203{,}973}{\text{Diameter in inches}}$$

Thus, a 72-inch propeller may be turned at 2,800 rpm maximum.

The typical lightplane wooden propeller weighs about 11 pounds. Each blade weighs about five pounds. At 2,750 rpm, each blade may have a total centrifugal pull in excess of 20,000 pounds acting on its root section. A metal fixed-pitch prop of the same size will weigh almost twice as much, and the forces trying to pull it apart increase correspondingly.

This centrifugal force actually has a beneficial effect. The rotating blades are subject to a strong twisting force due to the fact that no two points on each blade have the same velocity or the same pitch (angle of attack). The very high centrifugal force on each blade counteracts the twisting force and tends to straighten out each blade.

There are other forces that act upon a rotating propeller, all of them complex, including gyroscopic loads, torque loads, thrust loads (which change with aircraft attitude), and vibration.

Appendix B

Certification and Operation of Amateur-Built Aircraft, AC 20-27C

1. Purpose: This Advisory Circular provides guidance and information relative to the airworthiness certification and operation of amateur-built aircraft.

2. Cancellations: a. AC 20-27B and AC 20-28A, dated 4/20/72, and 12/29/72, respectively.

3. Background: a. The FAA has received many requests from amateur aircraft builders for information relative to the construction, certification, and operation of amateur-built aircraft. This AC provides guidance concerning the building, certification, and operation of amateur-built aircraft of all types; defines the eligibility of nationally advertised kits; defines how much construction the amateur builder must do to have the aircraft eligible for airworthiness certification; and describes the FAA role in the certification process.

b. The FARs provide for the issuance of experimental certificates to permit the operation of amateur-built aircraft. FAR 21.191 defines an amateur-built aircraft as an aircraft, the major portion of which has been fabricated and assembled by person(s) who undertook the construction project solely for their education or recreation. The FAA has interpreted this rule to require that more than 50 percent of the aircraft must have been fabricated and assembled by the individual or group. Commercially produced components and parts which are normally purchased for use in aircraft may be used, including engines and engine accessories, propellers, tires, spring steel landing gear, main and tail rotor blades, rotor hubs, wheel and brake assemblies, forgings, castings, extrusions, and standard aircraft handware such as pulleys, bellcranks, rod ends, bearings, bolts, rivets, etc.

4. Definition: As used herein, the term "District Office" means the FAA General Aviation District Office (GADO), Flight Standards District Office (FSDO, pronounced "Fisdo"), or Manufacturing Inspection (MIDO) District* Office that will perform the airworthiness inspection and certification of an amateur-built aircraft.

5. FAA Inspection Criteria: a. In the past, the FAA has inspected amateur-built aircraft at several stages during the construction of the aircraft. These inspections are commonly referred to as pre-cover inspections. The FAA also inspected the aircraft upon completion (prior to the initial issuance of the Special Airworthiness Certificate necessary to show compliance with FAR 91.42), and again prior to the issuance of the Unlimited Special Airworthiness Certificate. In the interest of streamlining operations within the government, and utilizing FAA inspection resources in areas most affecting safety, the FAA has reassessed its position concerning the need for all of these inspections. The following reflects the results of this assessment:

1. The amateur-built program was designed to permit any person to build and operate an aircraft solely for educational and recreational purposes. The FAA has always maintained that the amateur builder may select his/her own design and should not be inhibited by any overly stringent FAA requirements. The FAA does not approve these designs since it would not be practical to develop design standards for the multitude of unique design configurations that are generated by amateur builders.

2. FAA inspections of these aircraft are limited to ensuring the use of acceptable workmanship methods, techniques, and practices; verification of flight tests to ensure that the particular aircraft is controllable throughout its normal range of speeds and throughout all the maneuvers to be executed; and determination of operating limitations necessary to protect persons and property not involved in this activity.

3. In recent years, amateur builders have adopted a practice

---

*EMDO

whereby they call upon a person having expertise with aircraft construction techniques (such as Experimental Aircraft Association designees—reference paragraph 7d 1.), to inspect particular components; e.g., wing assemblies, fuselages, etc., prior to covering, and conduct other inspections as deemed necessary. This practice has been highly successful in ensuring construction integrity.

4. There are many instances where the FAA has found that pre-cover inspections were unnecessary, since in some cases the areas requiring inspection are readily accessible when the aircraft was completed. In other instances, pre-cover inspections were found to be neither meaningful nor feasible, such as in cases involving aircraft constructed from composite materials.

5. The FAA inspection previously performed after successful completion of the flight test, and prior to issuance of an Unlimited Certificate, was determined to be redundant in that any workmanship discrepancies would be detected during inspections performed prior to the issuance of the initial Special Airworthiness Certificate.

b. In view of the foregoing considerations, the FAA had concluded that safety objectives, relative to the amateur-built program, can best be accomplished, with no derogation of safety, by the use of the following criteria:

(1) Amateur builder should continue the practice of having knowledgeable persons (i.e., EAA Designees, FAA certified aircraft mechanics, etc.) preform pre-cover inspections and other inspections as appropriate. In addition, builders should document construction using photographs taken at appropriate times prior to covering. The photographs should clearly show methods of construction and quality of workmanship. Such photographic records should be included with the builder's log and other construction records.

(2) The FAA will conduct an inspection of the aircraft prior to the issuance of the initial Special Airworthiness Certificate to enable the applicant to show compliance with FAR 91.42(b). This inspection will include a review of the information required by FAR 21.193, the aircraft builder's logbooks, and an examination of the completed aircraft to ensure that proper workmanship has been used in the construction of the aircraft. Appropriate operating limitations will also be prescribed at this time.

(3) Upon completion of the required flight test, the FAA will review the results of the tests accomplished, as recorded in the air-

craft logbook. Upon a determination that compliance has been shown with FAR 91.42(b), the FAA will issue the Unlimited Special Airworthiness Certificate with expanded operating limitations.

6. Certification Steps: The following procedures are in the general order to be followed in the certification process:

a. *Initial Step.* The prospective builder should contact the nearest District Office to discuss his/her plans for building the aircraft with an FAA inspector.* During this contact, the type of aircraft, its complexity and/or materials should be discussed. The FAA will provide this prospective builder with any guidance necessary to ensure a better understanding of applicable regulations.

b. *Registration.* Prior to completion of the aircraft, the builder may apply for an identification number and register the aircraft. Detailed procedures are in paragraph 7A. This must be done before submitting an Application for Airworthiness Certificate, FAA Form 8130-6, under FAR 21.173.

c. *Marking.* The identification number (N-number) assigned to the aircraft and an identification plate must be affixed in accordance with FAR 21.182 and FAR Part 45, "Identification and Registration Marking." Detailed procedures are in paragraph 7b.

d. *Application.* The builder may apply for an Experimental Certificate by submitting the following documents and data to the nearest District Office:

(1) Application for Airworthiness Certificate, FAA Form 8130-6.

(2) Enough data (such as photographs) to identify the aircraft.

(3) An Aircraft Registration Certificate, Form 8030-3, or the pink copy of the Aircraft Registration Application, 8050-1.

(4) A statement setting forth the purpose for which the aircraft is to be used; i.e., "operating an amateur-built aircraft," FAR 21.191(g).

(5) A notarized statement that the applicant fabricated and assembled the major portion (reference paragraph 7e) of the aircraft, for education or recreation, and has evidence to support the statement available to the FAA upon request.

(6) Evidence of inspections, such as a logbook entry signed by the builder describing all inspections conducted during construction of the aircraft. This will substantiate that the construction has been

---

*Honest, this is the exact wording; but read it slowly and take the narrower meaning.

accomplished in accordance with acceptable workmanship methods, techniques, and practices, in addition to photographic documentation of construction details.

e. *Inspection and Issuance of Airworthiness Certificate.*

(1) After inspection of the documents and data submitted with the application, the FAA will inspect the aircraft, and upon a determination that the aircraft has been properly constructed will issue an Experimental Certificate together with operating limitations that specify the flight test area. The inspector will verify that all required markings are properly applied, including the following placard which must be displayed in the cabin or the cockpit at a location where it is in full view of all the occupants:

PASSENGER WARNING—THIS AIRCRAFT IS AMATEUR-BUILT AND DOES NOT COMPLY WITH FEDERAL SAFETY REGULATIONS FOR STANDARD AIRCRAFT

(2) Details concerning flight test areas are contained in paragraph 7c. The operating limitations are a part of the Experimental Certificate and must be displayed with the certificate when the aircraft is operated. It is the responsibility of the pilot to conduct all flights in accordance with the operating limitations, as well as the General Operating and Flight Rules in FAR Part 91.

(3) Upon satisfactory completion of operations in the assigned test area, the owner of the aircraft may apply to the FAA for amended operating limitations by submitting to the nearest District Office an Application for Airworthiness Certificate, FAA Form 8130-6. Prior to issuance of the amended limitations, the FAA inspector will review the applicant's fight log to determine whether corrective actions have been taken on any problems encountered during the testing and that aircraft condition for safe operation has been established.

7. General Information.

a. *Aircraft Registration.* The FARs require that all U.S. civil aircraft be registered before an Airworthiness Certificate can be issued. FAR Part 47, "Aircraft Registration," prescribes the requirements for registering civil aircraft. The basic procedure is as follows:

(1) Before an amateur builder can register his/her aircraft he must first obtain an identification number which will eventually be displayed on the aircraft. It is not necessary to obtain an identification number at an early stage in the project, especially if the builder intends to obtain a special number of his choice. Under FAR Part 47, a special identification number costs $10.00 and may be reserved for no longer than one year. Although this reservation does not apply to numbers assigned at random by the Aircraft Registry, the best time to request identification number assignments in either case is in the later stages of construction.

(2) To apply for either a random identification number or a special number, the owner of an amateur-built aircraft must provide information required by the FAA Aircraft Registry by completing an Affidavit of Ownership or Amateur-Built Aircraft, AC Form 8050-88. The affidavit establishes the ownership of the aircraft. If the aircraft is built from a kit, the builder should also submit a signed bill of sale from the manufacturer of the kit as evidence of ownership. Any communication sent to the FAA concerning aircraft registration should be mailed to the FAA Aircraft Registry, Department of Transportation, P.O. Box 25504, Oklahoma City, Oklahoma 73125.

(3) After receipt of the applicant's letter requesting identification number assignment, the FAA Aircraft Registry will send a form letter giving the number assignment, a blank Aircraft Registration Application, AC Form 8050-1, and other registration information. All instructions must be carefully compiled with to prevent return of the application and delay in the registration process. The Aircraft Registration Application while original and green copy should be returned to the FAA Aircraft Registry within 90 days prior to the estimated completion date of the amateur-built aircraft, accompanied by a fee of $5.00 payable by check or money order to the Federal Aviation Administration. The pink copy of the application should be retained by the applicant to be carried in the aircraft as a temporary registration until the Certificate of Registration is issued.

b. *Identification Marking.* When a builder applies for an Airworthiness Certificate, he must show, in accordance with FAR 21.182, that his aircraft bears identification and registration markings required by FAR Part 45. The following summary of the pertinent rules is provided for guidance.

(1) The aircraft must be identified by some means of a fireproof plate that is etched, stamped, engraved, or marked by some other approved fireproof marking as required by FAR 45.11. The identification plate must include the information required by FAR 45.13.
(2) The identification plate must be secured in such a manner that it will not likely be defaced or removed during normal service, or lost or destroyed in an accident. It must be secured to the aircraft in an accessible location near an entrance, except that if it is legible to a person on the ground it may be located externally on the fuselage near the tail surfaces.
(3) The builder's name on the data plate must be the amateur builder, not the designer, plans producer, or kit manufacturer. The aircraft model is whatever the builder wishes to assign, provided it will not be confused with other commercially-built aircraft model designations.
(4) The required registration marks on most amateur-built airplanes must be displayed on either the vertical tail surfaces or the sides of the fuselage. However, the builder should refer to FAR 45.25, which defines specific requirements for the location of marks on fixed-wing aircraft. The location of registration marks for rotorcraft, airships, and balloons is specified in FAR 45.27. These registration marks must be painted on or affixed by any other means ensuring a similar degree of permanence. Decals are also acceptable. The use of tape which can be peeled off, or water-soluble paint such as poster paint, is not considered acceptable.
(5) Most fixed-wing amateur-built aircraft are eligible to display registration marks with a minimum height of three (3) inches. However, if the maximum cruising speed of the aircraft exceeds 180 knots (207 mph), registration marks at least twelve (12) inches high must be displayed. The builder should refer to FAR 45.29, which defines the minimum size and proportions for registration marks on all types of aircraft.
(6) The identification marks displayed on the aircraft must consist of the Roman capital letter "N" (denoting U.S. registration) followed by the registration number of the aircraft. Any suffix letter used in the marking must also be a Roman capital letter. In addition, the word "Experimental" must be displayed on the aircraft near each entrance to the cabin or cockpit in letters not less than two (2) inches or more than six (6) inches in height.
(7) If the configuration of the aircraft prevents the aircraft from being marked in compliance with any of the above requirements, the builder should contact the FAA regarding approval of a differ-

ent marking procedure. It is highly recommended that any questions regarding registration marking be discussed and resolved with the local FAA inspector before the marks are affixed to the aircraft.

c. *Flight Test Areas.*

(1) Amateur-built airplanes and rotorcraft will initially be limited to operation within an assigned flight test area for at least 25 hours when a type certificated (FAA-approved) engine/propeller combination is installed, or 40 hours when an uncertified (not FAA-approved) engine/propeller combination is installed. Amateur-built gliders, balloons, dirigibles, and ultralight vehicles built from kits evaluated by the FAA and found eligible to meet requirements of FAR 21.191(g), for which original airworthiness certification is sought will be limited to operation within an assigned test flight area for at least 10 hours of satisfactory operation, including at least five (5) takeoffs and landings.

(2) The desired flight test area should be requested by the applicant, and if found acceptable by the FAA inspector, will be approved and so specified in the Operating Limitations. It will usually encompass the area within a twenty-five (25) statute-mile radius from the aircraft's home base. The FAA will ensure that the area selected is not over densely populated areas or in congested airways so that the flight testing, during which passengers may not be carried, would not likely impose any hazard to persons or property not involved in this activity. The shape of the flight test area selected may need to be modified to satisfy these requirements.

(3) The carrying of passengers or other crew members will not be permitted unless they are necessary to the conduct of the flight test while the aircraft is restricted to the flight test area.

(4) When it is shown in accordance with FAR 91.42(b) that the aircraft is controllable throughout its normal range of speeds and all maneuvers to be executed, and has no hazardous operating characteristics or design features, and the time period in the flight test area has been completed, the owner may apply for operation outside the assigned flight test area.

d. *Design and Construction.*

(1) Many individuals who desire to build their own aircraft have little or no experience with respect to aeronautical practices, workmanship, or design. An excellent source of advice in such matters is the Experimental Aircraft Association (EAA) located in Oshkosh, Wisconsin. The EAA is an organization established for the pur-

pose of promoting amateur building and giving technical advice and assistance to its members. The EAA has implemented a "Designee" program, the aim of which is to ensure the safety and dependability of amateur-built aircraft. Most EAA designees are willing to inspect amateur-built aircraft projects and offer constructive advice regarding workmanship and design. The FAA strongly encourages the use of such designees.

(2) Any choice of engines, propellers, wheels, and other components, and any choice of materials, may be used in the construction of an amateur-built aircraft. However, it is recommended that FAA-approved components and established aircraft quality material be used, especially in fabricating parts constituting the primary structure, such as wing spars, critical attachment fittings, and fuselage structural members. Inferior materials, or materials whose identity cannot be established, should not be used.

(3) The design of the cockpit or cabin of the aircraft should avoid, or provide for padding on, sharp corners or edges, protrusions, knobs, and similar objects which may cause injury to the pilot or passengers in the event of an accident.

(4) An engine installation should be such that adequate fuel is supplied to the engine in all anticipated flight attitudes. Also, a suitable means, consistent with the size and complexity of the aircraft, should be provided to reduce fire hazards wherever possible, including a firewall between the engine compartment and the fuselage. When applicable, a system providing for carburetor heat should also be provided to minimize the possibility of carburetor icing.

(5) Much additional information and guidance concerning acceptable fabrication and assembly methods, techniques, and practices are provided in FAA Advisory Circular AC 43.13-1A, "Acceptable Methods, Techniques, and Practices—Aircraft Inspection & Repair," available from the Government Printing Office, or TAB Book #2377.

(6) In the areas of engineering design, the builder should obtain the services of a qualified aeronautical engineer, or consult the seller of purchased plans or construction kits, as appropriate.

e. *Construction Kits.*

(1) Construction kits containing raw materials and some prefabricated components may be used in building an amateur-built aircraft; however, aircraft which are assembled from kits composed of completely finished prefabricated components and parts, and

precut, predrilled materials, are not considered to be eligible for certification as amateur-built aircraft, since the major portion of the aircraft would not have been "fabricated and assembled" by the builder.

(2) All aircraft built from a kit may be eligible for certification, provided that the major portion (more than 50 percent) has been "fabricated and assembled" by the amateur builder. The major portion of such a kit may consist of raw stock such as lengths of wood, tubing, extrusions, etc., which may have been cut to an appropriate length. A certain quantity of prefabricated parts such as heat-treated ribs, bulkheads, or complex forms made from sheet metal, fiberglass, or polystyrene, would also be acceptable, provided it still meets the major portion of "fabrication and assembly" requirement, and the amateur builder satisfies the FAA inspector that completion of the aircraft is not merely an assembly operation.

(3) Various material/parts kits for the construction of aircraft are available nationally for use by amateur aircraft builders. Advertisements tend to be somewhat vague and may be misleading as to whether a kit is eligible for amateur-built certification. It is not advisable to order a kit prior to verifying with an FAA inspector that the aircraft, upon completion, would be eligible for certification as amateur-built under existing rules and established policy.

(4) It should be noted that the FAA does not certify aircraft kits or approve kit manufacturers; however, the FAA does perform evaluations of kits that have potential for national sales interest, but only for the purpose of determining whether an aircraft built from the kit can meet the "major portion" criteria.

f. *Safety Precaution Recommendations.*

(1) All Aircraft.

(a) The pilot should thoroughly familiarize himself with the ground handling characteristics of the aircraft by conducting taxi tests before attempting flight operations.

(b) Before the first flight of an amateur-built aircraft, the pilot should take precautions to ensure that emergency equipment and personnel are readily available in the event of an accident.

(c)Violent (aerobatic) maneuvers should not be attempted until sufficient flight experience has been gained to establish that the aircraft is satisfactorily controllable throughout its normal ranges of speeds and maneuvers. Those maneuvers successfully demonstrated while in the test area may continue to be permitted by the FAA when the operating limitations are expanded.

g. *Documentation Requirements.*

(1) To preclude any problems or questions concerning source or specification of materials, parts, etc., used in fabricating the aircraft, it would be helpful if the builder kept copies of all invoices or other shipping documents.

(2) A construction log maintained by the builder for the project, including photographs taken as major components are completed will be acceptable substantiation that the builder constructed the major portion of the aircraft.

(3) The aircraft's flight history should be recorded in an Aircraft Logbook. The nature, as well as duration, of each flight should be documented. If the aircraft is considered acrobatic, the acrobatic maneuvers should be demonstrated in the flight test area and recorded in the Aircraft Logbook.

8. Recurrent Airworthiness Certification: When an amateur-built aircraft has completed operations in the assigned test area, it is eligible for an unlimited duration airworthiness certificate. With the issuance of the Unlimited Airworthiness Certificate, an operating limitation requiring a condition inspection at twelve (12) month intervals is imposed. The aircraft builder can be certified as a repairman to enable him to perform the condition inspection. Specific information regarding the repairman certification can be found in AC 65-23, "Certification of Repairman" (Experimental Aircraft Builders).

Index

Index

W